REFRAMING DEFORESTATION

International concern over the extent and rate of tropical deforestation has intensified, whether for interests of biodiversity, climate change, forest peoples or respect for nature. West Africa is assumed to have experienced the most dramatic and recent deforestation of all, supporting cataclysmic climatic predictions. Yet evidence presented in *Reframing Deforestation* suggests that the scale of destruction wrought by West African farmers during the twentieth century has been vastly exaggerated and that global analyses have unfairly stigmatised them and obscured their more sustainable, even landscape-enriching practices.

The book begins by reviewing how West African deforestation is represented in policy, forestry, and environmental sciences, and the types of evidence which inform present deforestation orthodoxy. On a country-by-country basis (covering Sierra Leone, Liberia, Côte d'Ivoire, Ghana, Togo and Benin), and using historical and social anthropological evidence, subsequent chapters evaluate this orthodoxy critically. Each country exemplifies different debates which have occurred in relation to each country's deforestation. Together the cases build up a variety of arguments which serve to reframe forest history and question how and why deforestation has been exaggerated throughout West Africa, setting the analysis in its institutional and social context.

Stressing that dominant policy approaches in forestry and conservation require major rethinking worldwide, *Reframing Deforestation* illustrates that more realistic assessments of forest cover change, and more respectful attention to local knowledge and practices are necessary bases for effective and appropriate environmental policies.

James Fairhead is Lecturer in the Department of Anthropology and Sociology at the School of Oriental and African Studies, University of London and **Melissa Leach** is a Fellow of the Institute of Development Studies, University of Sussex.

GLOBAL ENVIRONMENTAL CHANGE SERIES

The *Global Environmental Change Series*, published in association with the ESRC Global Environmental Change Programme, emphasises the way that human aspirations, choices and everyday behaviour influence changes in the global environment. In the aftermath of UNCED and Agenda 21, this series helps crystallise the contribution of social science thinking to global change and explores the impact of global changes on the development of social sciences.

Also available in the series:

ARGUMENT IN THE GREENHOUSE
The international economics of controlling global warming
Edited by Nick Mabey, Stephen Hall, Clare Smith and Sujata Gupta

ENVIRONMENTAL CHANGE IN SOUTH-EAST ASIA
People, politics and sustainable development
Edited by Michael Parnwell and Raymond Bryant

THE ENVIRONMENT AND INTERNATIONAL RELATIONS
Edited by John Vogler and Mark Imber

POLITICS OF CLIMATE CHANGE
A European perspective
Edited by Timothy O'Riordan and Jill Jäger

GLOBAL WARMING AND ENGERGY DEMAND
Edited by Terry Barker, Paul Ekins and Nick Johnstone

SOCIAL THEORY AND THE GLOBAL ENVIRONMENT
Edited by Michael Redclift and Ted Benton

ENVIRONMENTALISM AND THE MASS MEDIA
The North–South divide
Graham Chapman, Kevin Kumar, Caroline Fraser and Ivor Gaber

REFRAMING DEFORESTATION

Global analyses and local realities: studies in West Africa

James Fairhead and Melissa Leach

Global Environmental Change Programme

London and New York

First published 1998
by Routledge
11 New Fetter Lane, London EC4P 4EE

Transferred to Digital Printing 2003

Simultaneously published in the USA and Canada
by Routledge
29 West 35th Street, New York, NY 10001

Typeset in Garamond by
J&L Composition Ltd, Filey, North Yorkshire

British Library Cataloguing in Publication Data
A catalogue record for this book is available from the British Library

Library of Congress Cataloging in Publication Data
Fairhead, James, 1962–
Reframing deforestation: global analyses and local realities: studies in West Africa/James Fairhead and Melissa Leach.
p. cm. – (Global environmental change series)
Includes bibliographical references and index.
1. Deforestation – Africa, West.
I. Leach, Melissa. II. Title. III. Series.
SD418.3.A358F35 1998
333.75′137′0966–dc21 97–35302
CIP

ISBN 0–415–18590–4 (hbk)
ISBN 0–415–18591–2 (pbk)

Printed and bound by Antony Rowe Ltd, Eastbourne

To Gerald Leach, for his inspirational and critical scholarship in analysing environmental pasts and futures

CONTENTS

MAPS

MAPS

TABLES

ACKNOWLEDGEMENTS

This book is the product of our joint and equal co-authorship. It was made possible by a generous Global Environmental Change Research Fellowship to James Fairhead from the Economic and Social Research Council (ESRC) of Great Britain (No. L32027313393), and by programme support to Melissa Leach at the Institute of Development Studies from the then Overseas Development Administration (ODA). Its theoretical perspective, the research agenda and the joint nature of the analysis are rooted in our previous research in the Republic of Guinea (Fairhead and Leach 1996a).

We would like to thank the many people who have commented on our arguments and case material as presented in conferences and seminars during the research. In particular, we are grateful to those who have commented in detail on earlier written drafts, including Seth Afikorah-Danquah, Sara Berry, Reginald Cline-Cole, Christopher Fyfe, Edwin Gyasi, William Hawthorne, Eric Lambin, Gerald Leach, James Mayers, James McCann, Achim von Oppen, Paul Richards, Simon Rietbergen and Elizabeth Tonkin, and to the anonymous reviewers of the original manuscript. Responsibility for errors of fact and interpretation, to which such a wide-ranging text is surely vulnerable, remains, of course, ours alone.

While many of the works cited here are in French, we have translated all citations for consistency and to assist the English-speaking reader. We apologise if our attempts to convey their sense accurately have sometimes resulted in considerable loss of linguistic elegance from the original.

INTRODUCTION

According to FAO (1993) statistics Africa has the second largest annual deforestation rate among the world's continents, and West Africa has experienced the most rapid deforestation of all. In international conservation and forestry circles, and in discussions of global environmental change, West Africa's total loss of forest cover is portrayed as both dramatic and recent. The area of forest lost west of the Dahomey Gap during the twentieth century has been estimated as 30 million ha. As van Rompaey (1993: 13) recently summarised: 'Only about 8 million ha of West African forest remained in the mid-eighties. This is some 20 per cent of the precolonial area.' An even more dramatic analysis is presented by the World Conservation Monitoring Centre which suggests that of an 'original' West African forest cover only 13 per cent remains (Sayer *et al.* 1992: 74).

Such widely cited figures, albeit elaborated in different ways from place to place, frame analysis of, and assumptions about, the history of vegetation change and its causes; causes variously linked to population growth, to migration, or to socio-economic developments in commerce or technology. In this way, they also serve to frame more general understandings of regional society, population and history, of how West Africa's rural inhabitants live, and of how life has changed. They frame programmes and policies designed to redress forest loss in local, national and international arenas.

A concern with rapid and recent forest loss has, indeed, been the driving force behind forestry and conservation policies – and has contributed to more general strategies for rural development – throughout West Africa's forest zone. Since early colonial times the resulting programmes have had significant practical effects on rural people's lives. For instance, reserves have been established in which local use rights have been eroded, often totally. Governments have taken control over 'protected' tree species, whether on farms or elsewhere, and policies to regulate and suppress practices thought to be forest damaging, such as shifting cultivation and fire-setting, have imposed fines, and at times imprisonment, on resource users. Such policies have aimed, in various ways, at protecting 'nature'; but in doing so they have frequently denied resource control to inhabitants and have created difficulties for local livelihoods. Deforestation analyses have also underlain a variety of interventions to plant trees and establish woodlots, aimed in effect at replacing or rehabilitating a degraded 'nature'. Dominant views of deforestation have also orientated other rural development activities, whether the siting of roads and rural infrastructure or the orientation of agricultural activities, towards less forest-damaging forms.

This book argues that such dominant visions fundamentally misrepresent the history of West African forests, and the relationship between those forests and people. In particular, we will argue that the extent of forest loss during the twentieth century has been vastly exaggerated. Much so-called deforestation either took place much earlier, or has not taken place at all since the areas in question have not carried forest in historical times. Furthermore, we will argue that much of the forest that has been lost during the twentieth century, and indeed much of that which remains, covered land which had earlier been populated and farmed, calling into question the commonplace view of population growth and deforestation as linked one-way processes. We will argue that more people does not necessarily mean less forest, as is commonly assumed. Especially on the northern margins, it has on the contrary sometimes been people, their settlement and their land use which have been responsible for the development of forest vegetation where it was previously lacking. If our argument is correct, then dominant policy approaches in forestry and conservation require some major rethinking.

Vast areas of forest have, clearly, been lost for a variety of linked reasons – farming, logging, fire, mining, industrial plantations; processes which have been examined at length elsewhere, and which are not the prime focus of this work. Rather, we argue that while the area of forest lost is indeed vast, it may be only about a third of that suggested in the international literature. Equally, while the rate of deforestation might remain a concern, it has also been hugely exaggerated. Moreover, the types of forest which have been lost are not necessarily those highlighted in international and national literatures.

Our reframing of West African forest history derives principally from the analysis of historical data. As we shall show and explain, dominant views of forest loss and the science which informs them have made little use of historical sources. They have relied instead on methodologies which deduce the nature of past vegetation from observations of present landscapes: methodologies which have become widely accepted in the disciplines empowered to reflect on vegetation change and which inform policies to address it. Yet there is now a wealth of historical information about West African landscapes, sufficient, we would argue, to track the major changes this century, as well as new historical methods for examining longer-term climate and vegetation history. Acknowledging the caution with which such historical sources must be approached, the strategy of this book is to draw them into dialogue with existing assertions about forest cover change, and to explore the contradictions which emerge.

One key contradiction is between the views which have dominated in conservation circles concerning the impact of land use on forests, and the ways in which people have actually worked with ecology. Attention to historical sources reveals important ways in which inhabitants have, at times, enriched soils and vegetation and increased tree cover. In so doing, it often confirms inhabitants' own perspectives and landscape interpretations, which orthodox assertions had invalidated.

We also explore contradictions between the analyses of social and population history which had evolved in keeping with dominant views of forest loss, and historical evidence of forest cover change. In many cases, reframing forest history forces a reappraisal of theories concerning social and demographic history which have long dominated in West Africa. In cases, we draw on certain analyses by social and economic historians, showing how uneasily they sit with dominant views of forest loss, but how they support – and are supported by – a reappraised view of forest cover change.

In this process, we reveal techniques and practices in local land use and management which have been hitherto overlooked or underestimated in their landscape impact. Indeed, attention to historical data throws into focus important ecological processes linked to these techniques, but overlooked in ecological investigation to date. In so doing, our analysis is, we think, able to highlight new avenues for interventions in forestry and natural resource management.

Historical data also help to shed light on present forest composition and condition, whether in the forest zone proper or in its transition zone to savanna. In some cases, historical sources help to provide interpretations of features which ecologists have found puzzling; in other cases, they provide quite different interpretations of features about which ecologists and foresters have felt certain. The same indicators which are taken to suggest forest loss, such as patches of 'relic' forest in savanna or 'relic' trees in fields, might, for instance, be reinterpreted as indicators of forest advance.

In its attention to tracking vegetation history, this book contributes to recent developments in forest ecology. While dominant approaches in this subject have not been historically grounded, there have long been dissenting voices which have recently grown in volume and influence in line with what has been termed 'new ecology' (e.g. Botkin 1990; Worster 1993, 1994). Emphasising ecological variability over different timescales and the extent to which vegetation reflects unique patterns of interaction between variables, and thus emphasising the particularity of pathways of development over time, such forest ecology is fundamentally based on historical understanding. We show how the landscapes of the forest zone have emerged as the visual, physical manifestation of accumulated conjunctures of such ecological processes, interacting with the equally conjunctural social and demographic history of the region. Building up landscape history in this way provides a strong critique of much of the ecological and social analysis of the region, based as it is on unsubstantiated assumptions about vegetational pasts. The latter approach, we would argue, is no longer tenable.

Out of the analysis, we attempt to provide a more precise picture of the nature, extent and rate of forest cover change in West Africa. This is important not only for resource assessment and policy at the national level, but also for the modelling of vegetation and climate interactions at regional and global levels. To reiterate, while suggesting in particular that the extent and rate of forest loss have been exaggerated, we certainly do not deny that deforestation has occurred in certain places and at certain times. Nor do we deny that in certain areas, land users have faced and are facing livelihood problems linked to changing vegetation, albeit mediated by social, economic and institutional factors. We do, however, suggest that more precision concerning what has been lost, when and where is essential for a more precise identification of any problems, and hence for the formulation of effective policies to address them. Equally, we are not suggesting that all assertions of deforestation are untrue. What we do argue is that sufficient can be proved false to underline the need for all to be subject to more critical analysis than they have yet received.

Places, definitions and issues

Our region of analysis is the zone where moist and semi-deciduous forests are found west of the so-called Dahomey Gap, thus including the forest regions of Sierra Leone, Liberia, Côte d'Ivoire, Ghana, Togo and Benin (see Map I.1). This covers what is known as the Upper Guinea forest zone, which ecologists distinguish in terms of composition from the forests east of the Dahomey Gap which extend from Nigeria into Cameroon and beyond. In each country, we include the forest–savanna transition zone, typically associated with a mosaic of forest patches in savanna. In the focus on tropical forests in West Africa, the study complements and contributes to the desertification debate which has framed analysis of vegetation change in the drier regions to the north. Equally, in raising questions concerning the evidence for forest cover change and its relationship with policy, this work clearly speaks to broader interests in tropical forests in other regions of the world. Indeed, the conclusions we draw from West African evidence raise disquieting questions for forest analysis elsewhere.

Precisely what, however, do we mean by 'forest'? As we explore in Chapter 1, analyses of forest cover change are beset by problems of definition, and much forest is lost or gained in the translation between different definitions. One aim of this analysis is to identify instances where academics or policy-makers have been misled (or have misled others) by such

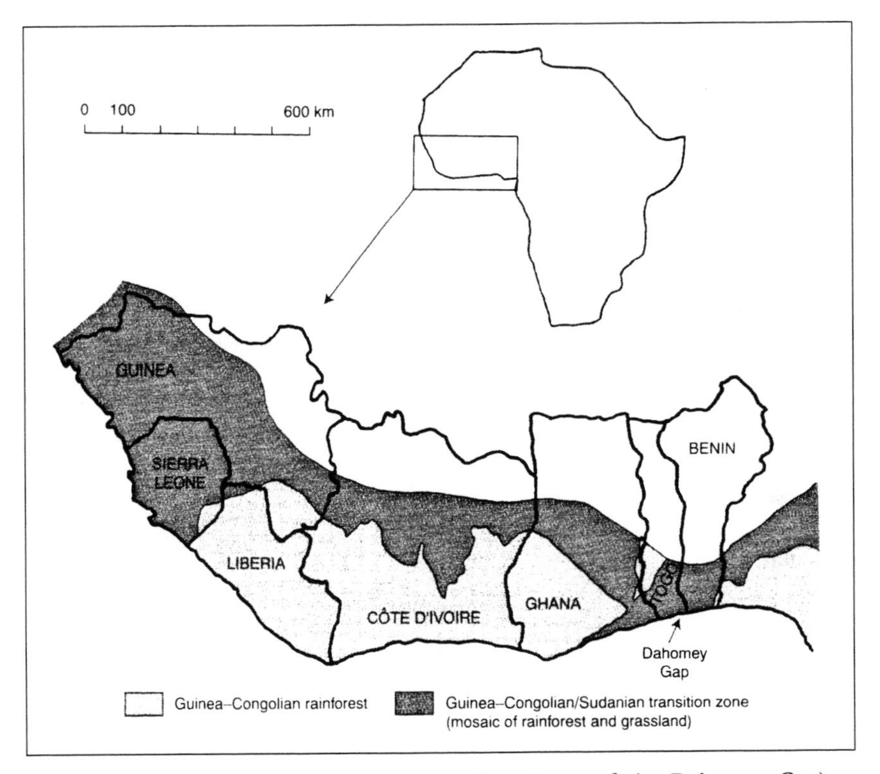

Map I.1 Humid forest regions in West Africa (west of the Dahomey Gap) as delineated by White (1983).

'definitional deforestation' (or, indeed, afforestation). Nevertheless, a general set of characteristics can be established which applies throughout the book's discussion unless otherwise specified. In keeping with the definition given by Hall (1987: 33), we consider forest broadly as 'vegetation dominated by trees, without a grassy or weedy under-storey, and which has not recently been farmed'. Forest thus defined is readily distinguishable from savannas which contain grasses, and from agricultural lands. We also circumscribe our focused concern to the evergreen and semi-deciduous forests of West Africa's humid and semi-humid zones, excluding the dry forests of the Sudanian woodland, savanna and Sahelian zones.

While it may be possible to gain precision concerning what vegetation consisted of and how it has changed, assessment of vegetation quality is a far more subjective exercise. Different people value vegetation qualities in different ways and for different reasons, and one person's degradation may be another's improvement. In this book, we address vegetation change mainly on the terms in which foresters and conservationists see it, and their 'objective' assessments of vegetation quality. But we nevertheless acknowledge that land users frequently have quite different priorities; for instance, preferring bush fallow for farming and collecting non-timber products over timber-rich high forest, and preferring a diversity of ecologies over ubiquitous high forest. This begs many questions concerning inhabitants' own diverse valuation and shaping of vegetation, an issue which we have explored directly elsewhere (Fairhead and Leach 1996a; Leach 1994) but which this book deals with only tangentially.

The forest region of Guinea is not included in this work as we have made it the subject of more detailed investigation elsewhere. Indeed, it was a series of studies in Guinea (Fairhead and Leach 1994a, 1996a) which provided the initiative for the critical perspective taken in the present work. As we elaborate in Chapter 1, a number of lines of critical enquiry emerged in the findings of our historical and social anthropological research there which prompted the present comparative investigation.

Strategy of analysis and presentation

This book examines forest history on a country-by-country basis. Rather different debates have occurred in relation to each country, and in this way, we feel it is possible to engage with them on their own terms, and to address the particular ways that forest science has articulated with social science around specific issues within each country. While there is inevitably some overlap between the concerns in each country our chapters are not reiterations of the same themes. Rather, we have sequenced them and the issues they address so as to build up in a narrative way the variety of arguments which together will serve to reframe forest history. So it is this explanatory concern, rather than geographical logic, which has guided our chapter sequence.

The first two chapters (Côte d'Ivoire and Liberia) highlight how the use of statistics and interpretations of remote sensing data can produce vastly exaggerated assessments of forest cover loss. Both also begin to hint at other themes which are developed in subseqent chapters: first (Liberia and Ghana), how analyses of demographic change have accompanied and mutually framed assessments of deforestation; second (Benin), how many of the vegetation forms which ecologists have been taking to indicate forest loss can be reinterpreted as indicating landscape enrichment by people; third, how the past existence and decline of forest is premised on historical analysis of socio-economic and technical change, notably in iron smelting (Togo) and the timber trade (Sierra Leone). As the chapters progress, so evidence accumulates to suggest an expansion of forest cover into savanna on the northern margins of the forest region under the complementary influences of land use and recent climatic change. Equally, many assumed 'pristine' forest areas prove, again and again, to be 'new forests' on land once densely settled and farmed, having grown following depopulation.

So we begin with Côte d'Ivoire because it perhaps best exemplifies problems with exaggerated forest statistics. The country has been accredited with the highest rate of deforestation in the tropics. Examining the evidence on which this case rests, we reveal major contradictions between modern representations of Ivorian vegetation at the turn of the century, and the observations of those present at the time. In this sense Côte d'Ivoire typifies the statistical rescheduling of forest loss to recent times by pushing forward the baseline date at which forest cover is assumed to have been more or less intact; an issue which recurs in subsequent chapters. Tantalising ecological and historical data suggest that savannas penetrated far further into the Ivorian forest zone in recent centuries, but were subsequently lost to forest.

A different, but equally controversial, story emerges from Liberia in Chapter 3, but here it is the data collected by early foresters which prove to undermine later forestry and social historical analysis. By many, Liberia has been seen as a case par excellence of a country wholly covered by little-disturbed forest until succumbing to the relentless, one-way pressure of a population and farming frontier. Many authors suggest that the country's forest cover has halved during the present century. Yet early foresters suggested that forest cover had been increasing during the nineteenth and early twentieth centuries, and furthermore that forest cover around 1950 was little different from that today. Ideas of unilineal forest loss may have framed much social and demographic analysis, but have simultaneously obscured what historians have recently identified as the far more complex and turbulent economic and political past of a 'worn-out country of great antiquity' (Mayer 1951: 25).

The case of Ghana combines the critiques raised in both Côte d'Ivoire and Liberia. It suggests that the statistics for forest loss in general circulation today massively exaggerate deforestation during the twentieth century, partly through assumptions that the bio-climatic forest zone consisted of intact forest around 1900. While many foresters and ecologists within Ghana know this to be untrue, the persistence of these statistics reveals the power of representation in international circles. Equally, as we examine in some detail, there is considerable evidence for long-term expansion of forest into savanna along the northern and eastern margins of the forest zone. As in Liberia, assumptions of relentless forest decline backdated to earlier centuries do not match Ghana's documented history of significant population decline in certain periods and places, frequently associated with the slave trade, and the linked appearance of 'new forests'.

The case of Benin in Chapter 5 shows that much of what today's analysts consider as degraded forest, characterised by oil palm, bush fallow and isolated forest patches and trees, may well indicate the opposite: the capacity of inhabitants to enrich their landscapes. While many of Benin's vegetation forms – as indeed similar forms in other countries – have been taken as evidence of recent forest loss and savannisation, there is strong evidence that in many cases people established them. As the case for deforestation in Benin has rested so strongly on deducing the deforestation process from these vegetation forms, these alternative interpretations suggest that a fundamental reinterpretation of the country's vegetation history is possible.

Togo, in contrast with the countries considered in earlier chapters, has rarely been seen as the subject of extensive deforestation in the twentieth century. Rather, deforestation has been pushed back to the nineteenth century and before, and attributed principally to the fuel requirements of the flourishing iron-smelting industry in the pre-colonial period. While seductive in analytical appeal, assumptions concerning the deforestation effect of the industry are questionable. While speculation concerning Togo's early vegetation may continue, of more fundamental significance to analyses of recent vegetation change in West Africa are findings that Togo's farmers have been establishing secondary forest thicket in savannas, contrary to dominant views that this formation is a degraded form of original forest. When these are coupled with similar findings in Benin, Guinea, Côte d'Ivoire and Ghana, a fundamental challenge is mounted to the view that African farming serves only to reduce and savannise forest cover.

Sierra Leone, the final country case considered, raises many of these earlier themes. Moreover, it exemplifies particular ambiguity in definitions of the 'forest zone', and the ways in which analysts can take mere redefinitions as evidence for actual forest cover change. This issue, combined with the interpretation of indicators as in Benin, has led some analysts to assume that Sierra Leone has undergone dramatic forest loss since 1900. However, more critical analysts have suggested that the country's major forest loss occurred during the early nineteenth century, as in Togo, but in this case linked to exploitation for timber exports and agricultural expansion. In this respect, Sierra Leone has been taken as a precursor to similar processes supposed to have occurred more recently in other West African countries. Yet a critical dissection of the cases for timber-led and farming-led deforestation in nineteenth-century Sierra Leone shows them to be wanting. If there was little forest following the timber exploitation period, it may have been because there was little there in the first place.

While the country chapters highlight different issues, then, they also contain recurring themes and common undercurrents which at once reveal the orthodoxy, albeit moulded into uneven relief. The cross-country analysis, furthermore, shows the weakness of the evidence on which, in all cases, analyses of forest history have rested. These assorted forms of evidence are reviewed in Chapter 1 and the suggestion made there that they provide a shaky foundation for analysis is amply borne out by the country case study findings. Indeed, in our ability to make these cases we in fact indicate how little is known of African social and economic history, or how little of what is known has secure foundation.

In a work of West African scope it has not been possible to subject each locality within each country to exhaustive detailed analysis. Indeed, the sources we draw on are generally documentary, consisting in the main of published travel, military, agricultural and political accounts, complemented by early cartography and some archival material. Where we draw on fieldwork, oral accounts and comparative analysis of air photographs, they are those published by other researchers and available to us as secondary sources; sources which provide detailed information for particular cases. Within this comparative and critical analysis, such examples, we would argue, raise fundamental questions which need to be addressed (if only to be rejected) for those other areas where we have been unable to gather such detailed historical information, or which have not been studied. In drawing on secondary sources, we fully acknowledge that some of the points in our critique have already been made by other researchers. However, these studies have generally remained marginal to debates concerning regional vegetation change, usually by circumscribing their claims to the very particular locality in question. By bringing these studies together into a comparative framework, this book shows that they highlight issues germane right across the forest region, and should not be minimised in the way that they have been hitherto.

A cautious approach to the accuracy of historical sources is essential in a work of this kind. Descriptions of vegetation early in the twentieth century were made for various reasons by observers who differed in their outlook, objectives and assessment methods. As we show in each country case, an awareness of these differences is important in evaluating the accuracy of particular estimates and their assorted biases. The largest figures for early forest cover generally derive from crude assessments of the area of 'forest', as distinct from savanna, according to early vegetation maps. In drawing up such maps, geographers such as Breschin (1902) and botanists such as Chevalier (e.g. 1912a) sought to give a broad-brush image of West African vegetation and forest potential without distinguishing between high forest, bush fallow and farmland. Information compiled in country maps almost always included a description and approximate placing of the forest–savanna boundary, because of its significance for military and agricultural reasons (e.g. Map I.2). These works are more useful for understanding the position of the forest–savanna boundary – albeit crudely, overlooking the ambiguity of the forest–savanna transition zone – than they are for their estimates of forest cover. This point was itself made in other early accounts, more focused on actual than on potential forest resources, such as those by the colonial botanists and foresters assigned missions to assess forest resources, whether timber, rubber, oil palm products or unused farmland. Colonial foresters such as Chevalier, Meniaud, Thompson and Unwin attempted to make detailed, on-the-ground and location-specific descriptions of vegetation status and quality.

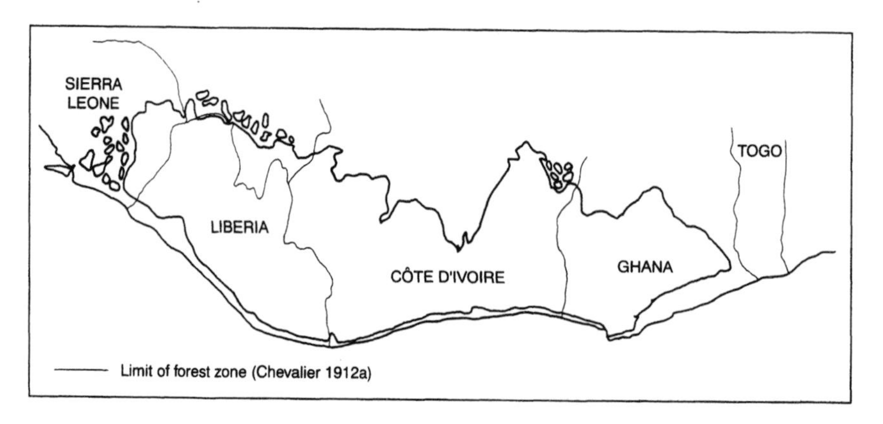

Map I.2 The humid forest zone and forest–savanna boundary as described in Chevalier's 1912 map of West African vegetation (Chevalier 1912a).

They depended for their data both on their own observations and on the reports of district administrators, which, by 1910, tended to be quite detailed. These descriptions, and the reports on which they are based, at their best constitute a rich source for site-specific forest history. There were inevitable potential biases towards more accessible sites: in particular those near roads and trade routes. But foresters themselves were aware of these problems and sought to overcome them in ways which often involved arduous travel through uninhabited country. Indeed, in many cases forest assessors were the first colonial administrators to visit reputedly uninhabited regions.

Early analysts often made crude estimates of national forest cover, whether or not solicited to do so by colonial administrations, which generally provide strong rejoinders to present-day estimates of past forest cover. In making their estimates, colonial foresters often combined descriptive sources with estimates of areas which were farmed, fallowed or under tree crops. The latter were derived from national agricultural statistics when possible. In other cases, foresters made their own crude calculations based on early figures available for prevailing population levels and the presumed area of farm/fallow land required per person. Again, these estimates must be approached with due caution, and clearly do not match the accuracy of modern estimates based on air photographic and satellite cover. Yet as guidelines of orders of magnitude, these figures are clearly more accurate than those based simply on estimates of the area of the forest zone, and were frequently generated in explicit criticism of the latter. Indeed, resource assessment missions often had to voice such criticism to counter more optimistic estimates of forest resources generated by elements within colonial administrations anxious to promote investment in their colonies. In many cases, these presented the forests as an inexhaustible resource, whereas the foresters and those who contracted them were at pains to stress resource limits, the climatic implications of forest cover loss, and the need for more 'rational' exploitation. Such motivations might suggest a tendency among foresters to underestimate forest cover area, and this can indeed be discerned in certain cases which we explore. That heated debates about the quality of estimates of forest cover took place at this date underlines the need for caution before taking any particular statistic at face value for understanding past vegetation cover.

Other sources that we draw on – descriptions by missionaries, 'explorers' and other travellers, and oral accounts – contain their own particular biases. For example, many travellers who came from drier regions emphasised the verdancy of the vegetation which they found in the forest zone, whereas those familiar with high rainforest described the same vegetation in a much less effusive way. Accounts of forest cover made from the ground are also notorious in overestimating the density of wooded vegetation because of the observer's low angle of vision. In the distance, farmland with scattered on-farm trees is hard to distinguish from dense forest. Vegetation history in inhabitants' oral accounts can be biased not only for these reasons, but also as a result of perceptions of vegetation immensity in childhood, and more significantly by the meanings of vegetation in local political discourse – where, depending on the cultural context, 'open savanna' or 'dense forest' may connote either uninhabited land, suggesting particular firstcomer status and tenurial claims, or earlier occupation by particular ethnic groups.

These problems in interpreting historical data limit the extent to which a definitive alternative vegetation history can be compiled, and the status of our findings. Nevertheless, problems in interpretation do not mean that early sources should not be used. As we argue, many present assertions concerning past forest cover are either not based on any historical sources, or make uncritical use of one or two. Other historical evidence simply falsifies them. In most cases, we would argue that it is possible to gain much better precision concerning forest cover at different dates by considering multiple sources together, and we therefore forward alternative forest cover estimates and accounts of forest history on this basis. At the least, the use of historical evidence should succeed in opening up a debate over competing interpretations of West Africa's ecological past, and in raising pressing questions which need to be addressed by future research.

Elements of our critique also build on themes already initiated in earlier works. A more respectful view of farmers' management of vegetation and ecology has been brought to recent audiences by Pélissier (1980), Richards (1985) and others attentive to the knowledge and practices of rural African farmers (see Scoones and Thompson 1994; Warren *et al.* 1995). Yet the implications of these works have not yet been directly counterposed to the literature dealing with deforestation. Equally, it is hardly original to suggest that much of today's tropical forest area, including that in West Africa, is not 'pristine' or part of a 'wild Africa', an observation made long ago and reiterated in a number of recent works (e.g. Adams and McShane 1992; Gomez-Pompa and Kaus 1992; McNeely 1994). Yet there has been little work which addresses the nature, extent and timescale of events which have influenced forest composition, at least in the West African region, and their implications for today's landscape, forest ecology and policy. In initiating this process, the concerns of the present study chime with those of 'new ecologists' who reject working assumptions of past forest equilibrium, instead adopting a more historically determined perspective on vegetation change. In placing people's roles in shaping forest landscapes centre-stage, this study joins a well-established and now rapidly growing body of work on landscape history in general (e.g. Sauer 1925; McCann 1995; Beinart 1996), and the socialised nature of Africa's anthropogenic landscapes in particular (e.g. Guyer and Richards 1996).

In Chapter 8 we examine how and why deforestation has been exaggerated throughout West Africa, by setting the scientific history of forest cover analysis in its institutional and social context. While analysts today may consider themselves to be making independent assessments of forest cover change, we suggest that their methods, theories, deductive reasoning, disciplinary authority and institutional affiliation reflect a colonially rooted genealogy from which present-day assertions cannot be separated. We therefore explore how a pan-West African orthodoxy concerning forest cover change developed in the early colonial period through the emerging science of forest ecology, the institutionalisation of its reasoning in scientific and administrative circles, and the alliances forged with colonial views of populations and their land use. We examine reasons for its persistence post-independence and up to the present. And we enquire how scientific ideas about forest cover change co-evolved with particular sorts of forest and conservation policy.

While this book reflects on the understandings of science and landscape which have come to inform forest and conservation policy, the scope of this work precludes a detailed account of the ways that particular policies were developed and put into effect. This would require attention to issues of administrative and political practice and expediencies in each country, in interaction with the diversity of local land management forms and interests. Such analysis requires the kind of focused depth that we have given it in Guinea (Fairhead and Leach 1996a), and as exemplified in works by a number of environmental historians, such as Grove (1994, 1995), Anderson (1984) and Beinart (1984). While this study raises general questions of policy relevance, then, this is at a more generic level, rather than at one which responds to particular policy concerns raised in each country.

Nevertheless, a broad analysis of West African forest history such as this has value in furthering policy debates, as we outline in the conclusion. While it suggests lines of enquiry for conservation policy on the ground, it also carries broader force to be felt in the international arenas in which statistics about deforestation circulate. Above all, it underlines the need for much greater analytical rigour and accountability among those who produce and promote figures with a remote, but nonetheless powerful, impact on the lives of Africa's farmers and forest inhabitants.

1

FORESTS OF STATISTICS

Deforestation in West Africa

> Droughts in West Africa over the last 20 years may have been caused by the destruction of rainforests in countries such as Nigeria, Ghana and Côte d'Ivoire. . . . Further deforestation in the region could cause the complete collapse of the West African monsoon. . . . At the beginning of this century, the West African coastal forests covered around 500,000 square kilometers. Since then, up to 90 per cent have disappeared to make way for farms and other kinds of human activity.
>
> (Pearce 1997: 15; see Zheng and Eltahir 1997)

Deforestation figures and their uses

According to analysis current in international conservation and forestry circles, West Africa has experienced dramatic forest loss during the twentieth century, accelerating during the last few decades. Ample statistics appear to bear this out, with region-wide figures – such as van Rompaey arguing that only 20 per cent of the pre-colonial forest area remained by the mid-1980s – apparently supported by those for particular countries. 'Lost Forests', as the *New Scientist* article cited above was headlined, might 'Leave West Africa Dry' (Pearce 1997: 15; see Zheng and Eltahir 1997).

Table 1.1 presents the recent FAO figures concerning the nature, extent and rate of forest loss over the decade timescale (1980–90), showing the relentless demise of remaining forest area with more than 10 per cent lost over the decade. Table 1.2 presents the data from an influential survey of anthropogenic vegetation change in West Africa during the twentieth century (Gornitz and NASA 1985), showing countries to have lost between 69 per cent and 96 per cent of the forest area which they had at the turn of the century; a loss graphically illustrated in Map 1.1. Table 1.3 presents the most recent statement concerning the extent of present forest in relation to the 'original forest' (Balmford and Leader-Williams in Sayer *et al.* 1992: 74), showing that except in Liberia, less than 13 per cent of the original forest cover remains. 'Original forest' is calculated with reference to the

Table 1.1 Forest cover change 1980–90 in West African countries according to FAO (1993)

Country	*Forest area (tropical rainforest and moist deciduous forest) (area, 000s ha)*	*Forest loss/year 1981–90 (000s ha)*	*Percentage of total forest lost/year*
Benin	4,183	56.7	1.4
Côte d'Ivoire	10,831	119.4	1.1
Ghana	9,151	134.0	1.5
Guinea	6,565	86.6	1.2
Liberia	4,634	25.4	0.54
Sierra Leone	1,889	12.3	0.6
Togo	1,318	21.8	1.6

Table 1.2 Anthropogenic deforestation during the present century (Gornitz and NASA 1985)

Country	*Forest area c. 1900*	*Present forest area (1985)*	*Percentage loss this century*
Benin	1,120,000	47,000	96
Côte d'Ivoire	14,500,000	3,993,000	72
Ghana	9,871,000	1,718,000	83
Liberia	6,475,000	2,000,000	69
Sierra Leone	not given	—	—
Togo	not given	—	—

Table 1.3 Present forest cover in relation to 'original' forest cover according to Sayer *et al.* (1992)

Country	*Original forest cover*	*Present forest area*	*Present forest as percentage of original*
Benin	1,680,000	42,400	2.5
Côte d'Ivoire	22,940,000	2,746,400	12.0
Ghana	14,500,000	1,584,200	10.9
Guinea	18,580,000	765,500	4.1
Liberia	9,600,000	4,123,800	43.0
Sierra Leone	7,170,000	506,400	7.1
Togo	1,800,000	136,000	7.6
Total	76,270,000	9,904,700	13.0

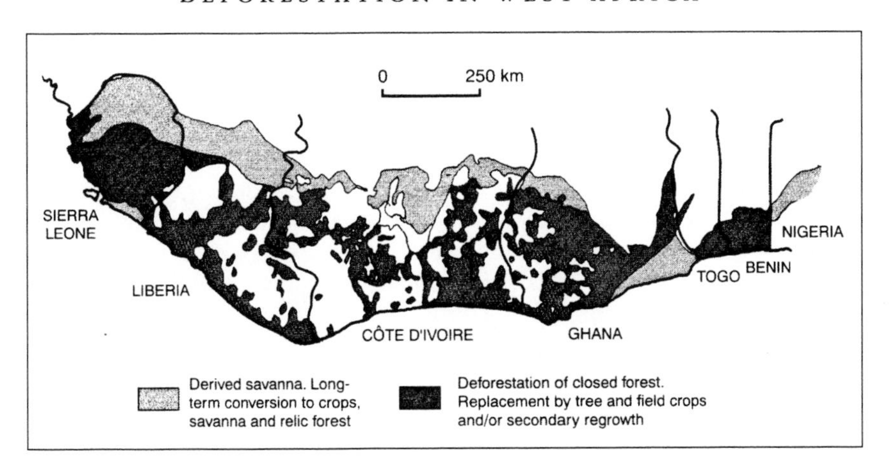

Map 1.1 Schematic map of anthropogenic vegetation changes in West Africa during the last century as described by Gornitz and NASA (1985).

vegetation map of White shown in the Introduction (Map I.1), incorporating all the forest–savanna mosaic of the Guinea–Congolia/Sudania transition zone as 'ex-forest'.

Such figures circulate widely in international organisations, where they are put to a variety of uses. They inform assessments of the extent of deforestation, important to the elaboration of international agreements as well as influencing the funding agendas of donor agencies and international conservation organisations. Equally the figures are central to the analysis of the regional and global consequences of deforestation, whether concerning biodiversity loss or global climate change. They can be fundamental in assessments of the causes of deforestation, when correlated with supposed causal variables, such as population growth and economic development. In each domain, the figures and assessments made on the basis of them are crucial to the formulation of the policies of international organisations such as the World Bank, the European Union (EU), the World Conservation Union (IUCN) and the World Wide Fund for Nature (WWF).

While as we shall see, national forest assessments have an ambiguous relationship with the figures in international circulation, in many circumstances government departments and non-governmental organisations are not averse to drawing on international figures when justifying funding for forest and conservation programmes, especially in front of international donors. Indeed, in drawing on national figures which feature in global tables, national assessments can perversely acquire greater authority and rhetorical weight. In particular, as Grainger notes, 'FAO's forest resource statistics are usually regarded as authoritative and so quickly become established in the literature by default' (1996: 73). Indeed, in the global literature they have been seen as a significant advance on earlier assessments

(Dixon *et al.* 1994). Furthermore, data from FAO, the World Resources Institute and other organisations are often the basis for unsourced tables and figures in secondary and tertiary articles and reports.

Assorted figures concerning deforestation have been incorporated into the scientific analysis informing international environmental agreements. The Inter-governmental Panel on Climate Change (IPCC), for instance, in assessing global carbon flux as related to tropical forests, relies heavily on the figures of FAO (1993), both directly and via other analyses which also draw on these figures (Brown 1993a; Brown *et al.* 1996; Dixon *et al.* 1994). The Global Environmental Facility, the World Bank funding envelope for combating biodiversity loss, also makes use of them. In particular, these figures are used in the construction of models which assess and predict global environmental change. Gornitz and NASA (1985), for instance, carried out their survey of twentieth-century deforestation specifically to identify albedo changes relevant to regional and global climate modelling (Henderson-Sellers and Gornitz 1984). Houghton (1991) and others argue that deforestation CO_2 emissions account for around 30 per cent of all anthropogenic emissions, a proportion that is growing quickly. During the 1980s forest-burning emissions expanded by 75 per cent, out of proportion to fossil fuel combustion. In the context of biodiversity, deforestation figures have been incorporated into models of rates of species extinction by using the species–area curves of MacArthur and Wilson (1967) (e.g. Raven 1987; Myers 1988; Wilson 1989).

Cross-country statistics concerning deforestation are also used in the analysis of causes of forest loss. By correlating country-by-country deforestation rates with other national variables such as poverty, income growth, external indebtedness, interest rates, land market values or structural adjustment, analysts attempt to identify the causal significance of each of these (e.g. Brown and Pearce 1994; Hyde *et al.* 1996). Indeed, as the discipline of environmental economics has gained strength, so international deforestation figures have increasingly been relied on to explore these economic variables as underlying causes. Similar methods are used to explore the impact of population growth, densities and agricultural land clearance (e.g. Allen and Barnes 1985; Burgess 1992). Others factors which are hard to explore using multi-country statistics – such as agricultural incentive policies, trade policies, and timber royalty and concession policies which underprice timber – have been explored in other ways, often based on single country cases, but which nevertheless depend on good figures concerning deforestation (see Repetto 1988, 1990; Barnes 1990; Sharma and Rowe 1992).

Environmental economics provides only the latest, and most model dependent, in a long line of social science analyses of deforestation causes. Many are less amenable to quantified assessment, but are nevertheless interlinked with a view of deforestation as recent and rapid, helping to explain it. With shifting cultivation usually seen as the key proximate cause, issues such as immigration into forest areas, poverty and tenure insecurity are highlighted in such analyses. Socio-economic and political changes over the past century, which may have compromised once effective 'traditional' systems of resource use and management, are seen as contributing to deforestation over the period; effects only further accentuated by population growth. The significance of these issues to analysts' views of modern rural society is clearly magnified, the more dramatic the deforestation with which they are linked. In this way, forestry and social science analyses can be seen to frame each other. This is exemplified in a recent World Bank overview of deforestation in West and Central Africa which argued that:

> [T]raditional farming and livestock husbandry practices, traditional dependency on wood for energy and for building material, traditional land tenure arrangements and traditional burdens on rural women worked well when population densities were low and population grew slowly. With the shock of extremely rapid population growth . . . these practices could not evolve fast enough. Thus they became the major source of forest destruction and degradation of the rural environment.
>
> (Cleaver 1992: 67)

Assumptions about the causes of deforestation have often been formalised in models designed to predict future deforestation rates and trends. IDIOM, for example, a model developed in the framework of the Tropenbos programme (Jepma 1995), was devised to simulate the effect of various policy options on deforestation rates, on the basis of changes in the global economy, timber supply and demand, and changes in agricultural land use. A second model, GEOMOD (Hall *et al.* 1995), is also designed to simulate rates and patterns of land use change, using spatial preferences in land colonisation (along rivers, up rivers and upslope) and preferences for adjacency and dispersion, considered on a region-by-region basis, to predict rates and patterns of tropical land use change. Supposed linkages between population growth and land cover change have formed the basis of numerous models predicting deforestation rates and associated carbon emissions (e.g. Houghton *et al.* 1993). For such models, deforestation figures are important for constructing the parameters of the model and calibrating them.

But as this work explores, those using data are often unaware of their basis and potential inaccuracy. Others have reservations about data quality, and acknowledge the need for more accurate assessments of forest cover change (e.g. Brown and Pearce 1994; Brown *et al.* 1996), but nevertheless use and recycle what is available. In this book the critical assessment of forest cover change in the humid forest region of West Africa can be taken as a test case, which may have more global relevance. It shows the extent of error inherent in these statistics, and the kind of dynamics to which improved assessments would need to pay attention.

The basis of forestry statistics

On what basis are assertions of the extent, timing and speed of deforestation made? The rest of this chapter considers some of the methods and deductions which have been most influential in generating the type of statistics employed in these assorted analyses. By outlining the forms of evidence and lines of argument which have guided analysis of deforestation to date, we highlight areas where they might be open to question, and hence avenues for their critique. This provides a broader perspective that contextualises our detailed country-by-country studies, which develop these critiques in relation to specific situations. Following the country-by-country critiques, and the questions they pose to such deforestation figures, we examine the history of these methods and their policy implications in relationship with colonial and modern institutions.

Observations at many scales and within many disciplines are implicated in the analysis of deforestation. Here it is convenient to begin with some of the key issues in the way international figures are derived. We move on to consider ways in which the national-level statistics which frequently inform international estimates are obtained, and then the ways that on-the-ground observations – and their interpretation within theories dominant in forest ecology – have been used to suggest deforestation. Many of the statistics of forest cover found in international conservationist literature are clearly borrowed from other reports of the same genre, thus circulating without returning to original sources. While many examples of this circulation are provided in later chapters, here we reflect on the bases from which original sources derive their estimates. We begin, however, with some definitional issues: how 'forest' is defined clearly affects estimates of change, and much forest can be lost or gained in the translation between different definitions. In approaching deforestation at different scales of analysis, very different definitions of forest are brought into play. It is partly in the interplay of such definitions that the critiques in later chapters find their foothold.

International and national forest interpretation

Definitional discrepancies in forestry analysis

Different perspectives on tropical forests lead to very different definitions of deforestation. It has become conventional to distinguish between restricted environmentalist definitions of forest, and more inclusive 'economistic' definitions (Barraclough and Ghimire 1990, 1995; Jepma 1995). Environmentalists, ecologists and conservation agencies such as WWF, IUCN and the World Conservation Monitoring Centre (Sayer *et al.* 1992) consider the impact of excessive logging, wood gathering, fire and livestock grazing as deforestation, degrading the forest ecosystem through loss of biomass and ecosystem services. Those defining forests in terms of economic forestry, such as FAO (1981), Lanly (1982) and the World Resources Institute (WRI), by contrast, tend to consider such processes as degradation, but not as deforestation unless they result in total conversion of forest to other land uses (see also Grainger 1993: 46). This more inclusive definition is exemplified in FAO, which defines forests as 'ecosystems with a minimum of [only] 10% crown cover of trees . . . generally associated with wild flora, fauna and natural soil conditions, and *not subject to agricultural practices*' (our emphasis) (1993: 10).

More restricted definitions of forest clearly give much lower estimates of forest area, and higher estimates of deforestation both in area (as logging can count as deforestation) and in percentage per annum (as it is relative to a smaller area).

Nevertheless, this broad definitional distinction cannot account, in West Africa, for the massive discrepancies between certain forest cover estimates; discrepancies which rest on inconsistent categorisation of forest types. FAO (1993), for example, suggests that in 1990 there was more 'forest' in Ghana than the area Ghanaian scholars consider to lie within the forest zone. This is equally true for Benin and Togo. This discrepancy is accounted for by the way FAO (1993) has defined 'moist deciduous forest'. The category includes not only vegetation usually considered in West Africa as tropical moist forest (i.e. moist evergreen forest, semi-deciduous forest) but also large tracts of Guinean savanna, usually classified as open savanna woodland, which fulfil the >10 per cent canopy cover criterion. Thus the FAO (1993) categorisation does not, infact, resolve the difference between forest and savanna (see Grainger 1996).

In contrast with FAO, analyses made in particular West African countries, as we will see, tend to adopt definitions of forest which exclude savanna and more closely parallel the more restricted definitions of ecologists and environmentalists. This type of definition is exemplified in Ghana by Hall (1987: 33), who defines forest as 'vegetation dominated by trees, without a grassy or weedy under-story, and which has not recently been farmed'. Forest thus defined is readily distinguishable from savannas which contain grasses. More subtle distinctions nevertheless exist within this genre of definition, based on a host of criteria, whether related to bio-climatic zone (as we discuss below) or to variations in canopy density. As we shall see, choice of criterion is frequently pragmatic, determined by the methods of cover assessment (i.e. satellite spectral resolutions, air photograph interpretation techniques), and this can lead to important discrepancies between assessments. Unless fine definitional distinctions are taken into account, especially when comparing cover estimates over short time intervals, trends in forest cover can be grossly misrepresented. Such discrepancies are best discussed in their specific contexts, and thus await analysis in the country chapters.

While FAO (1993: 10) infers that the land it classifies as forest is 'not subject to agricultural practices', the definition remains ambiguous in certain respects: whether various types of home gardening and tree cropping which leave high forest trees among tree crops are to be defined as forest, as well as the point at which fallow in regeneration comes to be classified as forest. The only way to account for the massive forest cover areas which the recent FAO figures present is, in fact, to assume that in practice areas of fallow and tree crops have been included. In this, FAO estimates again conflict with the way forest is usually defined within the countries in question to exclude land under recent fallow and tree crops. Unfortunately, as the FAO (1993) data have neither ecological nor policy relevance to West Africa, then, we have had largely to ignore them in our country-by-country analyses.

Remotely sensed imagery

With the growing availability of satellite data, it is often assumed that international figures for forest cover change – at least over the 1–2 decade timescale for which such data are available – are based on time series comparison of these. Indeed, many assumed that the assessment for FAO (1993) would be derived in this way. Instead, the FAO's analysis of forest cover change between 1980 and 1990 relied principally on previous national assessments. Only some of these were based on remote sensing data. Furthermore, the average survey year was 1980, and to create from these an estimate for both 1980 and 1990, forest areas usually measured only once (in 35 out of 40 African countries) were massaged using a mathematical model which explicitly linked national forest cover decline to rising population density. This not only depends upon assumptions about the relationship between population and forest cover change and about population history which we shall have cause to question, but also deflects uncertainty from statistics on vegetation cover to statistics on demographic change, which, many would argue, are even more notoriously suspect.

Even where time series satellite or air photographic data are available, they may not, in fact, provide the unproblematic source of accurate data on forest cover change which they first appear to. For any given moment in time, image resolution strongly influences what types of vegetation can be discerned, leading, almost by default, to forest being defined in certain ways. For instance, it is questionable whether satellite images and air photographs at a scale of 1:40,000 or smaller can resolve the distinction between forest and farmland with trees, whether long fallow, or forest canopy interspersed with planted tree crops (e.g. Wills 1962). Much satellite imagery is of extremely coarse resolution (notably NOAA/AVHRR) which fails to detect small-scale agricultural clearings, let alone disturbance through logging. Even higher-resolution imagery such as SPOT and LANDSAT may also be too coarse to map land cover, especially in low population density areas, where it fails to detect small, scattered fields (Wilkie 1994). The possibilities of manipulating spectral wavebands allow the interpreter of digital tapes considerable flexibility not only to match spectral mixes to resolve desired vegetation categories, but also to create images which bear little relation to vegetation on the ground. Such ambiguities can be partially resolved through careful ground truthing. However, this is not always carried out, and even when it is, important distinctions on the ground may be impossible to resolve in the remotely sensed imagery. Many modern interpreters are attempting to grapple with these problems, as both resolution and data manipulation become more sophisticated. Yet the need for such resolution and sophistication only serves to underscore problems in the results of earlier assessments which lacked these, and yet which appeared to provide clear pictures of vegetation cover.

Further potential problems arise in comparing remotely sensed imagery from different dates. While some seem naive, cases in this book show that they have often been overlooked. Comparison of images with different resolution – in effect, dealing with different categorisations of forest – can produce misleading results; a particular problem in comparing recent satellite data with older (e.g. pre-1960s) air photographs. The years themselves may be incomparable – for instance, in relation to rainfall levels and fire intensity; a problem reflected on, for example, by Grégoire and Gales (1988) in the methodology section of their analysis of 1975 and 1985 data in the forest zone of Guinea, but ignored in their conclusions. Images may also be taken at incomparable moments in the year; especially important in areas with pronounced dry seasons where vegetation has a more deciduous habit, and a tendency to burn annually. Images taken in the wet season or early in the dry season show a significantly higher canopy cover than those taken after fire has passed or trees have dropped their leaves. Misleading estimates of change do, of course, result should the date at which photographs were actually taken bear little relation to the date at which they are published, and taken up for use. In one set of estimates for Côte d'Ivoire, as we shall see, estimates of forest cover taken to be for 1966 were actually based on 1955–7 imagery, vastly distorting the rate of change of forest cover when these older images were compared with later ones. In some cases, as we shall examine especially in relation to Togo and Benin, regional and national figures which purport to be based on remotely sensed imagery may turn out to be extrapolated from relatively small sample sites where time series data happen to exist. Yet attention must be paid to why these sites were chosen, and whether they were sampled statistically in a way that could support their use in generalisation.

Notwithstanding the difficulties outlined above, remotely sensed imagery can give a relatively secure picture of vegetation change over a timescale of recent decades. In the countries considered, systematic air photography began in the 1940s or 1950s, and satellite imagery from the early 1970s. Generally, however, individual studies have not exploited the time depth of this coverage. Instead, in most cases in the areas we shall be examining, they have focused on change over a timescale of a decade. Not uncommonly, inferences from the results of such short-term comparisons are extrapolated backwards into the past. Thus, if forest loss over the past decade has been rapid, this is assumed to be merely the latest stage in an ongoing deforestation process throughout the past century.

These forms of imagery were, of course, not available to early colonial observers. Nevertheless, they did make estimations of national forest cover and of forest quality, in ways and under conditions which we outlined in the introduction. While these estimates must be approached with due caution, this does not mean that they should be as overlooked as they generally have been. As the country-by-country analyses will explore, these figures can provide insight into forest conditions early in the century, especially when used in conjunction with other sources concerning land use and forest quality.

Vegetation zones

Just as the notion of 'forest zone' was important to early calculations of forest cover, so it has been and remains important in providing a baseline against which to assess the extent of deforestation to date. Indeed, rather than compare present vegetation with the historical record of past vegetation, it has been common and expedient for studies dealing with a timescale of a century or more to compare current cover with the vegetation considered to be typical of a given bio-climatic zone; that is, assuming that the zone which could support a given vegetation once did support it. The 'zone' is thus taken to represent the baseline or 'original' vegetation. This is exemplified in Table 1.3 (p. 2), where original forest vegetation in each country is considered to be that of the zone thought able to support it. In many cases, and as we shall document in detail in the country-specific chapters, modern authors equate the integrity of this zone with the 'pre-colonial' period, implying that the forest was more or less intact c. 1900. Others use estimates of historical populations and assumptions about population–land clearance ratios to model by how much land cover differed from original vegetation at different dates since 1850 (Houghton *et al.* 1983, 1995), suggesting that 'large-scale conversion of forests to cropland and pastures occurred . . . still later [than 1940] in tropical Africa' (Houghton 1992, cited in Brown *et al.* 1996: 77). Both this reasoning and this dating, as we shall suggest in later chapters, have been responsible for introducing huge errors into West

Africa's deforestation statistics. The origins of this reasoning, its rise to dominance and its impact on policy and popular consciousness are discussed in Chapter 8.

At the scale which informs most national and international statistics, bioclimatic zones are divided into the forest zone of the higher-rainfall coastal belt, ceding to drier forest forms and then savannas further north. At a finer scale, ecologists have subdivided the forest zone according to several criteria: gradients which relate spatial trends in species composition to trends in rainfall, relief and edaphic factors (e.g. van Rompaey 1993); variations in deciduous or evergreen habit and in the presence or absence of grasses (Aubréville 1949; Hall and Swaine 1976, 1981); and in relation to characteristic species associations (e.g. Taylor 1960; Guillaumet and Adjanohoun 1971; Hall and Swaine 1976, 1981).

Many studies of forest cover change assume that at origin, these bioclimatic zones supported their 'climax' vegetation, i.e. the ultimate stage in the succession through which vegetation would progress in the absence of disturbance. In lay terms, this is the 'natural', 'virgin', 'pristine' or 'primary' vegetation of each zone. Where the vegetation does not accord with that deemed appropriate to the prevailing bioclimatic conditions, it is commonly construed as a degraded form of the former. Generally the disturbance is attributed to people, so that vegetation is an anthropogenic derivate and, if relatively stable, a sub-climax. This reasoning encapsulates a particular set of theoretical tenets within ecological science as it has been applied to forests. While this has been strongly challenged in the ecological literature of temperate and dry tropical zones, it remains the orthodox view among those dealing with tropical humid forests, especially in West Africa. Even though certain ecologists working on West Africa have come to reflect critically on this reasoning (e.g. Hawthorne 1996), it has remained unchallenged as a central plank in national and international forest analysis, and their ideas of baseline vegetation. The cases which we will present further challenge these tenets for the forests of West Africa, suggesting that the history of human and bioclimatic impacts renders the concept of 'baseline vegetation' highly problematic even over the century to millennium timescale.

In conceptualising anthropogenic influence on forest vegetation, analysts find it convenient to distinguish several types of conversion. The first is from high forest to various forms of secondary forest and forest thicket, as fallows developed following forest clearance for farming. In many circumstances, such forest thicket is regularly recleared for farming (every 4–14 years), in which case the land can be considered to be held within a rotational bush fallow cycle which might revert or 'escape' back to high forest if farming were abandoned. A second type of conversion is from forest to various forms of permanent farming, whether perennial tree crops or continuously cultivated annual crops. A third type of conversion is from forest or various types of bush fallow to savanna, as a result of grass invasion associated with repeated burning; a process commonly held to result either from an over-shortening of fallow periods consequent upon population or production pressures, or from farming at all in areas where forest is marginal. Such 'derived savannas', it is commonly argued, would return to their original forest in the absence of fire and people. These processes have been the subject of extensive categorisation and investigation in West Africa, including experiments to exclude fires from savanna and examine subsequent vegetation succession, and experiments to create savanna from forest using fire (e.g. Charter and Keay 1960; Ramsay and Rose Innes 1963; Adjanohoun and Assi 1968). The cases which we will present show this typology to be partial, capturing only processes by which people have converted forest to other forms, and failing to capture reverse dynamics.

While the overall concept of a forest zone remains dominant in analyses of original vegetation, its extent has been the subject of considerable debate. This is particularly so for the savannas lying on the northern margins of the forest zone. While one school of thought holds that this savanna is derived from forest, and thus properly belongs to the forest zone, another school is more cautious in attributing such origins. Often referring to the vegetation of this zone descriptively as forest–savanna mosaic, they would define it separately from the forest zone proper (e.g. White 1983). This variation in extent of forest zone and hence of supposed original forest, of course, affects estimates of the notional baseline area from which forest has been lost and consequently assessments of the extent and rate of that loss.

Equally fundamentally, the assumption that today's climatic zones once carried their characteristic vegetation depends on the assumption of climatic stability. In the most systematic description of original vegetation in relation to bioclimatic zones, Aubréville was explicit in dealing with what he calls 'the present climatic period which dates back certainly beyond the dawn of history' (Aubréville 1938: 77).

In the analysis of forest–climate relationships in West Africa to date, most emphasis has been placed on a very long timescale, with many authors noting the extreme arid phase pre-12,000 BP (Martin 1991). Nevertheless, they tend to construe this as a long-past episode, with 'present' climates having stabilised c. 3,000 years ago. Such a view is still upheld by Sayer *et al.* (1992), who, using the version of climatic history of Hamilton (1992), visualise forests as recovering after the Pleistocene refuge period, and then remaining stable until their recent degradation.

If climate has changed during more recent historical times, the change has conventionally been seen as due to deforestation, in line with long-dominant theories that deforestation engenders climatic desiccation. In raising the possibility that climates to support forest and forest itself might have disappeared together, this has made it possible to deduce that forests once existed in areas which are too dry for them today. Thus Aubréville, for example, calculated the old forest extent by inflating present rainfall figures by 15–20 per cent, the amount lost by deforestation according to trials in France, and then estimating past forest area with assumptions correlating this climate with forest potential (Aubréville 1938: 88). It is this kind of reasoning which has underlain the view that today's forests are subject to, and their composition responding to, progressively drier conditions; or, put another way, that vegetation zones are shifting southwards.

Nevertheless, Aubréville, later in his life, began to realise that today's forest vegetation seemed to be advancing into savanna zones, and concluded that vegetation today may still be recovering from the Pleistocene dry period, and that it had not yet reached equilibrium with current climate. This observation was incorporated into ecological debates deeper within the forest zone concerning whether or not forests reproduced their species composition, with Aubréville (1938) noting even at this date that they seemed not to (then explaining this in other ways) but with Swaine and Hall (1988) arguing that they seemed to. Appreciation that forest may still be responding to climatic rehumidification at a stroke undermines assumptions that today's climatic zones once carried an 'original' vegetation. Aubréville changed tack, interestingly as we shall see later, long after his enduring legacy concerning deforestation analysis had been inscribed in traditions of forest cover analysis in West Africa.

Subsequently, climate historians have come to recognise that there have been deep and long-period fluctuations in West Africa's climates in more recent historical times, incorporating broad phases of rehumidification as well as desiccation. Indeed, there is now growing evidence that it may be rehumidification from more recent times which is most significant to present vegetation dynamics. That there was a very dry phase in West African coastal climates around 3000 BP has been noted in several recent studies (in Gabon, see Schwartz 1992; in Cameroon, by Talbot 1981 and pers. comm., also Maley pers. comm. to Schwartz 1992). The importance of climatic change over this timescale to the forest zone has been underscored by Schwartz in Gabon:

> Palynological studies have shown that the climatic extension of open formations [c. 3000 BP] preceded the arrival of farmer–metallurgist populations usually considered to be very aggressive towards their environment . . . the present savannas are not of anthropic origin.
>
> (Schwartz 1992: 35; see also de Foresta 1990)

Moreover, other evidence exists that even more recent climatic changes may be relevant to the West African region, important over a timescale of centuries. That there have been more recent – if less profound – dry phases, including a recent dry phase c. 1300–1850, is suggested by Nicholson

(1979, 1980). Nicholson's historical evidence is supported by evidence from the levels of Lake Bosumtwi in Ghana, which were in decline or low from around 700 years ago until they began to rise c. 200 years ago, forcing lakeshore villages to be abandoned because of inundation (Talbot and Delibrias 1977). As Nicholson (1980) notes, descriptions and rainfall measurements in Freetown in the 1790s, for example, suggest that normal rainfall may then have been perhaps only half its present levels (see Winterbottom 1803). The potential importance of more recent climatic fluctuations is also evident in recent archaeological analysis of vegetation. For example, in the forest–savanna transition zone of Nigeria analysis of snail fauna dated from the first millennium suggests that, over a period of 300 years around this time, land was becoming progressively more forested, surprising the investigators, whose starting hypothesis had been the inverse (Barber 1985).

Notably, the work of Nicholson combining historical analysis and climatological investigation argues that climatic trends in the coastal forest zone may be the inverse of those in the Sahelian regions, so periods of Sahelian drought frequently and structurally correspond to moister periods near the coast (Nicholson 1980). Thus, the supposed effects of the well-publicised Sahelian droughts of recent decades should not be assumed onto West Africa's forest region. Equally, arguments which we will forward concerning climatic rehumidification and forest advance should not be generalised to West Africa's drier Sudanian and Sahelian regions.

Thus, the scale of the temporal dimension to West Africa's forest–climate relationship may be properly measurable not in thousands, but in hundreds of years. Brooks (1993), in a major historical work, has attempted to reframe West Africa's social and political history in this light, with a focus mainly on the drier regions and the period 1100–1600, although ignoring the inverted relationship between coastal and inland climates. The implications of climatic change over this timescale for vegetation and ecology have not been systematically analysed. It would not be inconceivable for today's vegetation to be responding simultaneously to recovery from drier conditions at each of these timescales – i.e. from protracted and deep aridity around 12,000, from a short, deep aridity 3,500 and from a relatively arid phase 700–200 years ago. Whether expressed through effects on soil, soil fauna and flora or vegetation distribution, lag effects from each dry and humid phase might remain relevant, interfering with present responses to more recent climatic variation (see Fairhead and Leach 1996b).

An appreciation of climatic variability, especially during historical times, and of time lags in vegetation response, clearly challenges the concept of baseline vegetation. Moreover, given the significance of climatic rehumidification as well as desiccation, it could be that spatial variation in present vegetation may be responding to this, or, in a more chaotic way, to climatic oscillation. The question then becomes the nature, direction and timescale of climatic changes about which evidence is rapidly accumulating. While we draw on evidence from climatic history to comprehend regional ecological changes, so the evidence from forest history which we begin to compile may itself be useful in the ongoing search to clarify the nature of West Africa's past climates.

Local landscape interpretation

The delimitation and characterisation of bio-climatic zones in forest analysis are complemented by observations of landscape on the ground. Botanists and foresters have used observations of particular vegetation forms, interpreted in the context of ecological theories about plant communities, to deduce vegetation history. It is to the role of such deductions in the assessment of forest cover change that we now turn. We cross-examine the assumptions underlying these historical deductions and question their validity. In this we draw comparatively on our earlier research in Guinea (Fairhead and Leach 1996a), which provided some quite different interpretations of features that ecologists have taken to indicate forest loss.

The Guinea research, using historical and ethnographic methods, showed how interpretation of landscape features in the forest–savanna mosaic of Kissidougou prefecture in terms of deforestation had obscured ways in which inhabitants had been successfully managing and enriching their landscapes. It indicated, for example, how the many patches of forest found in the savanna were not relics of a once extensive forest cover as had been thought, but owed their existence to inhabitants who had encouraged them to form around savanna settlements. It showed how secondary forest thicket had been established in grassy savannas and was not, therefore, degraded forest as had been thought. It showed how palms had been established in savannas, rather than being relics of the savannisation of forest. The contrast between policy assessments of vegetation change and the historical data for it could not have been starker. The same evidence which scientists and policy-makers had been taking to indicate vegetation degradation actually indicated anthropogenically induced regeneration and landscape enrichment (Leach and Fairhead 1994a, b, 1995; see also Fairhead and Leach 1995b, 1996a, b).

Ethnographic research in Kissidougou showed problems in interpreting each 'indicator' of forest loss. Each was better interpreted in other ways and, in sum, these alternative interpretations produced a very different, even reversed, reading of landscape change – a reading supported by other historical evidence from aerial photographs and archival documents. The study, then, raised questions concerning the validity of using conventional landscape interpretation methods in the assessment of deforestation, and showed the need to review the evidence for deforestation for countries and regions where these methods have been used to suggest it. This book takes on the task of reviewing historical evidence country by country. In the remainder of this chapter, we review at a more generic level how landscape features have been interpreted to suggest deforestation and savannisation, and identify certain deductions which might be open to question. We examine the deductions involved in assertions about two key processes of forest loss: the conversion of forest to savanna, and the conversion of high forest to bush fallow and farmland.

The arguments put forward by scientists and policy-makers concerning each of these types of conversion are rather different, but on the northern margins of the forest zone, the two issues are sometimes conflated into one process as forest is seen to be converted into bush fallow and thence to savanna. This chipping away of the forest at its northern margins is, indeed, one of the major convictions of deforestation analysis. As Nyerges once argued:

> [T]he history of the southern guinea savanna, in fact, is one of constant chipping away at the forest edge. . . . The zones of 'derived' savanna or forest–savanna mosaic frequently marked on vegetation maps reflect this process of savannisation, in which disturbed forest sites are invaded by grasses that subsequently burn and prevent or retard the regrowth of forest and the redevelopment of soil. Characteristics of this zone, which imply a history of degradation, include a sharp forest–savanna boundary, the presence of forest outliers and emergents in savanna, and the mosaic pattern of primary forest, secondary forest, farmland, and tall grass savanna that constitutes the transition between the forest and savanna zones.
>
> (Nyerges 1987: 327–8)

Forest conversion to savanna

As we have seen, the view that savannas have been derived from forest has been based heavily on the assumption that forest is the 'natural' vegetation of the forest–savanna mosaic and Guinea savanna zones. The calibration of vegetation appropriate to a bioclimate or soil type has depended on the existence of patches of vegetation, or vegetation forms, which are seen to be relics of the 'original' vegetation. Patches of dense humid forest in the savannas on the northern margins of West Africa's forest zone have been treated in this way. Frequently found as sacred groves or islands of high forest surrounding villages, in a landscape otherwise characterised by savanna, these sites are assumed to have been protected from the savannisation wrought elsewhere, being the least disturbed remnants of the natural vegetation. While foresters and botanists have commonly identified such 'remnants' from ground observation, interpreters of satellite imagery are frequently tempted to make similar deductions about forest–savanna dynamics from vertical snapshots of the forest–savanna mosaic landscape.

Deductions about long-term forest–savanna dynamics from the presence of forest patches have been reinforced by analysis of their species composition; whether the presence of particular species, or overall species diversity. When species considered typical of little-disturbed dense humid forest are found in such patches, it reinforces the view that these patches represent the zone's 'proper' vegetation. On the other hand, the presence in forest patches of species associated with drier zones is commonly taken as an indicator of processes of gradual desiccation; an indication that forest and climate are moving together towards a drier type. In effect the image is of vegetation zones shifting southwards with each taking on the characteristics of its more northerly, drier neighbour. The view of forest cover change therefore becomes an adjunct to arguments about desertification and the southwards expansion of the Saharan margins.

These deductions rest on the assumption that forest and forest–savanna dynamics are one-way processes. Yet West Africa's climate history renders such assumptions extremely doubtful. Even if forest (and forest climates) are disappearing together and giving way to savanna forms, this may not be for the first time. Forest areas may expand and contract and forest types change in various directions, in relation to climatic change. Foresters sometimes suggest that the presence of forest species in pockets of humid land many hundreds of kilometres from any other exemplars indicates that the intervening land once carried such forest, which has now been lost. The colonial forester Aubréville did so for Côte d'Ivoire, for instance, arguing that because the floristic composition of the forest outliers in the savannas of the Baoulé-V was almost identical to that of the principal mass, they must once have been joined to it: 'In the Baoulé, for example, certain [relict] are signalled only by some stands of high trees; silk-cotton (*Ceiba pentandra*) and *Cola cordifolia* above all, above a prairie of high grass' (Aubréville 1938: 80–81). But while it may indeed be the case that at one time such contiguous forest did exist, the presence of these relic species tells us nothing about intervening climatic and vegetation fluctuations. Relic species can only be an indicator of ongoing deforestation if one assumes unilineality, which cannot be upheld. Thus they tell us little or nothing about recent processes, at least in the way that they have been used. Assessing a timescale of vegetation change from such relic species would depend on having time-dependent data about the plant individuals concerned, perhaps based on genetic mutation rates.

Furthermore, the assumption that forest patches in savanna are relics of original vegetation ignores the possible role of anthropogenic factors in their origins, form and composition. In Kissidougou, patches which had been taken as relic forests turned out to have been enriched and altered in relation to both vegetation and soil fertility. The existence of Kissidougou's peri-village forest islands depended on the effects of settlement, soil enrichment through household waste disposal and gardening, and fire protection in creating peri-village conditions in which woody vegetation could establish. The form and species composition of these forests reflected their management and use as pre-twentieth century fortresses, as kola and coffee planting sites and as forest product reserves. Their composition was almost an archive of changing political and economic circumstances, while their location reflected population and settlement history. Similar forest patches existing without associated settlements were found to overlie past habitation sites, whether abandoned villages or farm hamlets (Fairhead and Leach 1996a). Certain species were found to be typical of these anthropogenic forest islands; for example, the silk-cotton trees (*Ceiba pentandra*) often established in rings around villages for fortification or fire protection.

Underlying Kissidougou's anthropogenic forest islands are equally anthropogenic soils. Analysts of forest–savanna dynamics and forest ecology have long recognised the importance of soil variability in comprehending vegetation distribution at all scales. Neverthless, arguments concerning anthropogenic soils and soil history (e.g. Anderson and Posey 1989; Balée 1989; Hecht and Posey 1989) have not been incorporated into these ecological debates. Many of the soil enrichment processes and related vegetational effects found to be important in Kissidougou, which we discuss in subsequent chapters, would, however, make good sense to soil scientists (see Fairhead and Leach 1996a). In highlighting these enrichment processes we certainly do not intend to deny the equally important dynamics of soil loss and impoverishment adding to soil variability, although these too perhaps ought to be considered in the more critical light in which we are examining ideas of forest degradation (see Stocking 1996). Nor would we want to suggest that all soils are equally malleable: in Kissidougou farmers themselves define certain soils which would not support forest even when anthropogenically enriched.

That the Kissidougou case may not be an exception in West Africa is suggested by other documented cases – outside the focal countries of this book – where vegetation has been similarly enriched around savanna villages and ruined villages. Keay described 'anthropogenic' forests which came to cover ruined towns in the savanna region of old Oyo in Nigeria, owing their existence to habitation (Keay 1947; see Abimbola 1964). Equally the *Kurmi* forest islands described in Nigeria by Lamb are thought to have been established around villages because of habitation:

> *Kurmis* [forest islands] due to the protection by man are nearly always small and are usually the result of fire protection of the land round a village that is situated at the side of a stream bearing fringing forest. . . . But for fire protection, these would have been a closed savanna woodland type, for which both soil

> and climate are suitable. Fire protection, however, has allowed high forest to become established. This type may be seen at Gwada and Tagbare, north of Minna, where the climate could support high forest but soil conditions are not good enough till altered by man's interference.
>
> (Lamb 1942: 188)

Similarly, anthropogenic forest islands have been described in Gabon by Aubréville, who obscures their significance by considering them an isolated exception:

> We observed a very curious case, incontestably the installation of forest by people in poorly wooded savanna of *Hymenocardia acida*. Batéké country (Middle Congo), near Okoyo and Evo, is scattered with islands of dense forest; at a glance one can see 10, 20 in the landscape. Each one marks the site of an old village. The Batéké always install their villages in savanna and cultivate their manioc only in savanna; in the rare vestiges of ancient forest which remain, they only plant a few yams, and some sugar cane; only the women cultivate, and they do not want to enter the forest. In the villages, to make their houses and fences the inhabitants use poles of iron-wood, *Milletia laurentii* . . . and *Ficus* sp. These trees propagate by cuttings very easily, the poles take root, oil palms spontaneously establish around houses, other species disseminated by seed subsequently establish. After a few years this forest vegetation becomes too dense for the taste of the Batéké, who do not like forest, and the village is moved a little further, in clean savanna. The old site transforms in a dozen years into a thick wood, mixed with palms. Whereas in general the populations of the forest zone seek forest for hiding in, living in and for defence, and clearing it for cultivation, the Batéké form an original exception, fleeing the forest, and re-creating it.
>
> (Aubréville 1949: 318; see Guillot 1980)

Other aspects of vegetation form in the forest–savanna mosaic have also been taken as indicators of forest retreat and savannisation. For instance, foresters and ecologists of West Africa have repeatedly considered the presence of oil palms (*Elaeis guineensis*) to indicate that original forest has retreated from the area, on the grounds that they hardly regenerate in regularly burned savanna. As Keay argues, for example, 'Oil palms are of special significance. It is well known that they readily regenerate in secondary forest but hardly at all in savanna. Oil palms in savanna are therefore a very good indication of former forest conditions' (Keay 1959b: 430; see also Allison 1962; Davies 1964). Yet in Kissidougou, where these assumptions dominated scientific and policy circles, oil palms turned out to have been spreading northwards into savanna landscapes over the past few decades, encouraged in this by farmers. Oil palms can be established in savannas in conjunction with farming, which provides the necessary fire protection during the early years. Their numbers tend to increase greatly in instances when such savannas are converted to bush fallows. Thus, it is very problematic to deduce recent deforestation from the presence of palms in savanna, and to assume that areas where oil palms thrive today have been forest in recent times.

Conversion of high forest to fallow

Treating high forest patches as indicators of a now-lost past vegetation is not confined to what is today the forest–savanna transition zone. Within inhabited areas of the forest zone, the land in the immediate vicinity of villages may also be high forest in an area which is otherwise bush fallow or farmland. It is frequently assumed that such patches date from a time of extensive, little-disturbed high forest cover. Yet evidence from Kissidougou and elsewhere suggests the possibility that high forest patches have been encouraged to grow up by people, within what was already bush fallow – and perhaps had been for centuries. More significantly, can the species composition and character of such patches be taken to reflect the region's original forest cover, given the altered fertility of their underlying soils, and the likely management of the species within them? Even in uninhabited forest, the densest, apparently oldest parts of the forest frequently overlie the ruins of settlement sites. Indeed, ecologists who are attracted to these areas by the pristine appearance of often massive trees, considering them to be the most representative of 'nature', have frequently usually been surprised to find pottery sherds in their soil pits (Jones 1956; van Rompaey 1993).

The bush fallow vegetation surrounding islands of forest in the forest zone is usually thought unequivocally to have been converted from high forest. In Kissidougou, such vegetation was described in the national forestry plan as 'post-forest' (Republic of Guinea 1988). But here, in a region towards the northern margins of the forest zone, such forest fallows proved to be a vegetation form which had converted from savanna within the last forty years. Further south, in the area between Kissidougou and Macenta, research based on nineteenth-century documents, oral accounts and air photographs drew the same conclusions (Fairhead and Leach 1994a). These other interpretive possibilities for bush fallow vegetation, which dominates large areas of the forest zone, at once cast doubt over one of the major premises on which analysis of forest loss has been based.

The presence of isolated tall trees of forest species in such bush fallows (or even in savannas) is commonly taken as unequivocal evidence of their high forest origins. When such trees have a long, straight, unbranched bole it is assumed that they grew up in a forest environment. But it is frequently the case that the tree species observed are those known to be 'self-pruning', and thus able to acquire this form irrespective of surrounding vegetation. Furthermore, there is evidence from Kissidougou and elsewhere that villagers 'train' such trees, whether inadvertently in lopping branches for fuel, or deliberately to increase light penetration to crops below, or in past times of warfare actually to encourage the development of lookout posts. Certain such species may be protected or indeed planted by farmers in their fields, and thus may grow to full forest height despite their location on open land.

Conclusion

In the Kissidougou case, the reinterpretation of landscape features which had been taken to indicate forest loss and savannisation contributed to a quite different reading of vegetation history; one which centred on inhabitants' activities in shaping and enriching the vegetation in the landscape with desired species and vegetation forms, in the context of climatic rehumidification over the century timescale. Paradoxically, what appeared to be the most undisturbed vegetation forms turned out to be the most disturbed. It cannot, of course, be assumed that these alternative interpretations hold for other parts of West Africa. But a major aim of the country-specific chapters which follow is to interrogate the extent to which they might, insofar as this can be done from secondary sources.

Given that the speculation of early observers concerning forest degradation was premised on these same assumptions, and given that such assumptions and deductions should properly be questioned, there is a clear need to be critical of the way that modern scholars accept early observers' analyses of change as scientific fact. Thus, in our analyses in the following chapters, while we make use of early colonial accounts for their descriptions of vegetation, we are also careful to assess their interpretations of prior forest history in a critical way.

The foundational methods and principles of analysis outlined in this chapter – at the scales both of international and national forest interpretation and of the interpretation of local landscapes – comprise the near totality of those employed in old and, as we shall see, in most modern analysis of West African forest cover change. Thus if these foundations can be shown to be flimsy, the whole edifice of deforestation analysis, together with its supportive statistics and linked social science analysis, begins to look extremely suspect.

2

CÔTE D'IVOIRE

Record rates of forest cover loss

Following FAO/UNEP's forest resource assessment for 1980 (FAO 1981), Côte d'Ivoire acquired the reputation of having the highest rate of deforestation in the tropics. Several authors capture the drama. Parren and de Graaf wrote recently that: 'Côte d'Ivoire has experienced the most rapid deforestation taking place in any country in the world since the mid-1950s. The average annual deforestation rate, as a percent of the remaining forest, rose from 2.4% in 1956–65 to 7.3% in 1981–85, over ten times the pan tropical average rate of 0.6%' (1995: 29). Gornitz and NASA (1985: 290–3) complement these deforestation rates with figures for forest cover. By 1980, they write, 'around 70% of the forest present in 1900 had been cleared. The area of closed forest in Ivory Coast decreased from 15 million ha in 1900 to 4.46 million ha in 1980, representing a decrease of 70.3%.' 'The forest was not exploited until 1880', write Arnaud and Sournia (1979: 290). 'Until 1951, exploitation was very limited. From 1951–7 it developed slowly . . . Extending over 12 million ha in 1956, the Ivorian dense forest now counts less than 4 million ha.'

The record-breaking statistics produced for Côte d'Ivoire make the case a particularly pertinent one with which to begin our critical analysis. In this chapter we focus principally on the generation and circulation of these statistics, and on how other data sources might call them into question. The Côte d'Ivoire case also begins to show how dominant views of deforestation have obscured counter-evidence concerning past populations and land use and the nature of past vegetation cover.

All recent authors examining Ivorian deforestation show remarkable consistency, largely because they draw on each other, or on the same original sources. Indeed, this orthodoxy was well summarised in the FAO/UNEP (1980) assessment whose publication drew comment from several other authors. Thus numerous articles appeared at around the same time, used the same data, and drew the same conclusions. They are summarised in Table 2.1, updated to 1990 with the influential forest cover assessment

Table 2.1 Dense humid forest cover change as represented by modern authors (millions of hectares)

Forest cover in:	*FAO 1981*	*Myers 1980, 1994*	*Arnaud and Sournia 1979*	*Monnier 1981*	*Thulet 1981*	*Bertrand 1983*	*Sayer et al. 1992*	*Parren and de Graaf 1995*	*Fair 1992*
1900	14.5	16.0			15.6	14.5		14.5	14–16
1955	11.8	11.8	12.0	11.8	11.8	11.8			
1965	9.0		9.0	9.0	9.0	8.9	8.6		
1973	6.2**		5.0	5.4*	5.4	6.2			
1977			4.0						
1980	4.0	3.62				3.9	4.4		
1990	2.7***	1.6					2.7	2.6	2.0
Total loss	11.8	14.4						11.9	12–14

Notes
* Extrapolation of partial air photographic cover made in 1972 (Sodefor 1975).
** Revised interpretation of results of Sodefor (1975).
*** Figure from Sayer *et al.* (1992).

produced recently by the World Conservation Monitoring Centre. These works continue to influence more recent assessments of Ivorian forest cover change (e.g. Fair 1992; Parren and de Graaf 1995).

From these figures, it appears that the bulk of forest loss has occurred during the twentieth century. Most analysts link this to a common set of causes:

- the introduction of cash crops (cocoa and coffee);
- immigration into the forest zone, population increase and food crop expansion there, assisted by
- logging roads, land clearance and the timber industry; and
- development projects assisting agricultural development at the expense of forests.

Several authors suggest a slightly deeper history. Monnier (1981), for example, argues that deforestation initially became significant as an effect of European contact in the sixteenth century. Previously, 'the man of the African forest' possessed only a small selection of food products, inconducive to the development of high populations, and was 'powerless' in front of the forest. He suggests that it was the importation of New World crops (especially maize and cassava) which underlay a nutritional and demographic shock so that 'the 16th and 17th centuries were a turning point in demographic tendencies' (1981: 180–1). Monnier dates the origin of settled farming to the sixteenth to eighteenth centuries within the northern margins of the forest, which, he argues, were savannised during this period. The population, caught between poor lands to the north and forest lands which simply could not be savannised to the south, over-exploited the transition zone and ruined its prior forest vegetation. Indeed, he argues, the location of and vegetation patterns within the forest–savanna boundary reflect the ravages of this early farmer-led deforestation and its limits in front of the imposing forest. Early populations thus revealed a 'latent boundary' (Monnier 1981: 181); a state of affairs which remained until the colonial period when cash crops stimulated further southward penetration into the forest zone. Lanly (1969) summarised a similar history: 'Until the 17th century the area of forest was not modified by the inhabitants who lived in close symbiosis with the natural environment. From the 17th century, and more particularly since the start of the colonial period, the introduction of food crops, export crops and population increase due to immigration' were responsible for a major reduction of forest area (1969: 46).

In this chapter, we will argue that recent authors have been massively exaggerating the extent of forest loss this century and that this is obscuring instances in which farmers have been managing their lands sustainably, or indeed enriching them. The exaggeration has been achieved principally by vastly inflating the extent of forest in the early to mid-twentieth century, as well as by some uncritical – and rather selective – use of air photographs. Furthermore, as we go on to argue, the social and demographic history imaged in these forestry works is highly misleading. Obscuring the deeper past, and Ivorian experience of it, is conducive neither to understanding the country's ecology, nor to the elaboration of acceptable (and workable) conservation strategies.

Twentieth-century deforestation: a critical look at the statistics

While there are many publications which examine Ivorian deforestation, most of these rely on a very limited number of sources. Indeed, most can be reduced to two key publications: Lanly (1969), which attempted to identify changes in forest area between 1955 and 1965; and FAO (1981), which used these figures, adding an estimation for 1980 based on them, updated using agricultural land use statistics, and an unsourced figure for 1900. Like many authors, the FAO analysis presumes that the whole area of the bio-climatic 'forest zone' was more or less intact forest around 1900. Thulet (1981: 153) is explicit on this point. Working for the forestry service on the National Environmental Commission, and author of 'The state of the environment in Côte d'Ivoire', he writes that 'In 1900, the area effectively covered by the forest corresponded almost to that of the forest zone: 15,600,000 m ha' (see Myers 1980; Parren and de Graaf 1995, who also make this assumption).

We argue here that neither the figure for 1900, nor those produced by Lanly (for 1955 and 1965 and hence for 1980), can be accepted. As a result all subsequent derivative analyses have serious shortcomings in their examination of past vegetation cover.

Air photograph interpretation

Lanly's 1969 study, which drew on a major CTFT evaluation of national timber potential (CTFT 1966), used maps based on the complete air photographic cover for Côte d'Ivoire in 1954–6 and a non-random sample of the air photographs to calculate the area of the forest zone and the percentage of dense forest within it. The study then calculated the forest cover for 1966 by comparing the proportion of land under forest in a sample of comparable 1956 and 1966 photographs. It concluded that forest cover had declined from 11.8 mha in 1955 to 9.0 mha in 1966.

But it seems hard to match the assertion that 74 per cent of Côte d'Ivoire's forest zone was forest in 1955 with observations made in the 1950s by the colonial foresters in Côte d'Ivoire, Aubréville and Mangenot, who knew the country well. They suggested that 'the reserves of the forest department represent the only almost intact specimens permitting study of the dense forest today' (Mangenot 1955: 6–7). He elaborated:

> [B]otanists searching for a fragment of 'black' (high) forest often have some difficulty finding them. These exist, however, in the form of isolated forests of greater or lesser size. Over the last 40 years the forestry service has reserved numerous of them. . . . Most of these relics are not large. Certain are still imposing, above all in the western region of the country between the Cavally and Sassandra rivers where appear to remain vast, really uninhabited regions.
> (Mangenot 1955: 21–2)

If forest reserves totalled about 4.5 mha, and forest in the exceptional south-western area about 2 mha, these together represented only 6.5 mha (40 per cent of the supposed forest zone). Mangenot may even have been exaggerating somewhat, as his comments here were intended to support the labours of Aubréville, who was responsible for most colonial forest reservation at the time. Nevertheless, Aubréville himself noted that the reserved forests were already separated from the broken forest around them (Aubréville 1957). When put on the spot, Aubréville estimated the forest cover of Côte d'Ivoire for 1950 to be 7 mha (Aubréville 1956), more or less half that suggested by the Lanly (1969) study for this time.

Mangenot was explicit about the illusions which the uncautious observer might hold about Ivorian forest cover:

> Seen from the summit of a hill . . . the landscape . . . is a sea of trees. . . . But when one flies over the region, or travels over it

> following the tracks, one sees that this corresponds, in reality, over vast areas with a corpse: the forest has been destroyed, with only a few large trees surviving in the shade of which are palm groves, coffee, cocoa, kola, plantains, and fields of cassava and yams. Each village is therefore at the centre of a zone not *dewooded* – large trees exist everywhere, and species cultivated are small trees, bushes and giant forbes – but *deforested*. A fragment of high forest has been replaced by a mosaic of plantations, fields and bush fallows, of small secondary woods.
>
> (Mangenot 1955: 21)

That the CTFT-derived study hugely exaggerated forest cover in 1955 is also suggested by other sources. A second analysis of Ivorian forest cover was conducted almost simultaneously in 1967 by Guillaumet and Adjanohoun (1971). Using ground truthing and stereoscopy, they claim to have been able to distinguish from forest a category of 'everything which relates to farming activities: annual or permanent fields, fallows however old, secondary forests (not ancient secondary forests), cocoa and coffee plantations' (1971: 164). FAO (1981) argues that this study confirms the CTFT analysis, as from its maps a forest cover of c. 8.8 mha can be inferred, similar to the CTFT estimate. Crucially, however, the forest cover map in Guillaumet and Adjanohoun (1971) was drawn up from the 1954–6 air photographs, not those of 1966. Thus, rather than supporting the CTFT figure of forest cover for 1966, Guillaumet and Adjanohoun's work refutes it, suggesting that it was in 1955 that Côte d'Ivoire had c. 8.8 mha, not 1966.[1]

That the CTFT results so influenced the FAO/UNEP assessment for Côte d'Ivoire is perhaps unsurprising since Lanly was director of the overall FAO/UNEP (1980) forest resources assessment. As Table 2.1 indicates, the potential errors in this study, compounded by the misinterpretation in the dating of Guillaumet and Adjanohoun's work, have left an enduring legacy for national and international analysis of Ivorian deforestation. Recently, for example, the World Conservation Monitoring Centre based its influential analysis of Ivorian deforestation rates on figures derived from Guillaumet and Adjanohoun's map, but again erroneously considered the map to have been derived from 1965 air photographs.

A further reason to question the CTFT estimates of Ivorian deforestation derives from early colonial assessments of forest cover. Certain early forest analysts did consider the majority of the Ivorian forest zone to be forest. Thus, for example, Zon and Sparhawk's global assessment of forest resources suggested that: 'The dense tropical forest starts at the coast, and almost without break covers more 12 mha' (1923: 868). In 1902, the geographer Breschin used the accounts of early French colonial 'explorers' to estimate a high forest cover of around 15 mha. These sources have often been adopted uncritically; for example, by Gornitz and NASA (1985), who cite Zon and Sparhawk to suggest that the Ivorian forest was 'intact' in 1900.

Nevertheless, these oft-cited calculations of forest cover were never accepted by the early French colonial forestry administration. Meniaud, head of the Ivorian forest service and concerned with both protection and production aspects of forests, was explicit. Under the heading 'Statistical errors concerning the surface of "Grand Forêt", the empty spaces in the interior of the extreme limits, and the reasons for these spaces', he suggested in 1933 that: 'The areas generally given in statistics as being occupied by high forest are calculated according to the extreme limits [that is] 11 million hectares for the high forest zone of Côte d'Ivoire.' By taking into account farmed and savanna areas in the forest zone, Meniaud calculated that 'the primary forest, or that exploited only by the export timber industry', then totalled only 8 mha (1933: 539). Indeed, certain large savanna inliers within the forest zone were clearly marked on early maps. Chevalier's 1912 map (see Map 2.1), for example, includes a vast savanna south-east of Dimbokro of

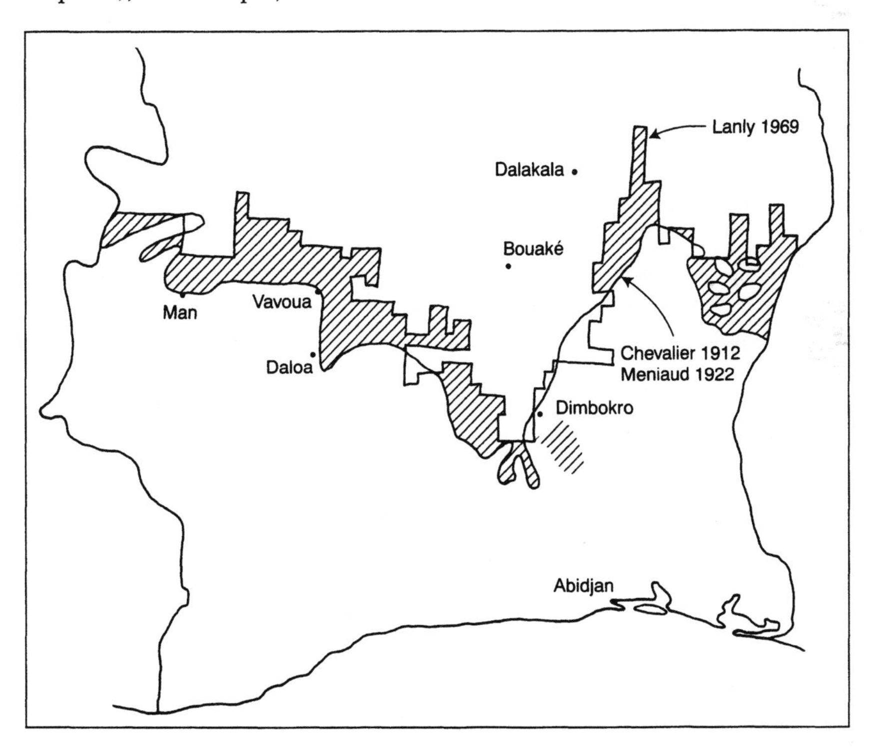

Map 2.1 Comparison of the northern limit of the Ivorian forest zone as described by Lanly (1969) and subsequent authors, with the limit as described in early vegetation maps (Chevalier 1912a; Meniaud 1922). The shaded area indicates the area now considered to lie within the forest zone, but left out of it early in the twentieth century.

around 100,000 ha.[2] And Aubréville noted the presence of open grasslands 34 km south of Daloa, far within the forest zone (1937: 241).

Meniaud's deflating estimate may contain an element of exaggeration, intending to image the forest as under threat. Yet it was, in fact, larger than those of his contemporaries, as Table 2.2 summarises. Indeed, Chevalier, who made two visits to Côte d'Ivoire, spending 1.5 years detailing its botany, put the figure of unused forest at only c. 6 mha (1909c). While such figures are of course problematic to interpret, and sometimes obscure in derivation if not 'off the cuff', they certainly cannot be taken to support the orthodox view that there were 14.5 mha of intact forest around 1900, gradually declining to 11 or even 9 mha by 1955. Indeed, they blatantly critique this view and the assumptions on which it is based. Were one to take them at face value – which should not be done – it is even possible to interpret the data to suggest an increase in forest cover in Côte d'Ivoire during the early years of the century. We might note in passing that the CTFT figure of forest remaining in 1955, at 11.8 mha, was equal to if not larger than estimates of the entire forest zone made between 1909 and 1940.

Several of these early authors, such as Gros, who was sent to make a formal evaluation of timber resources, gave precise descriptions of the regions of the forest zone where 'good' forest existed from a timber perspective (e.g. Gros 1910; Chevalier 1912b), as if to drive home the point that such forest was not to be found everwhere. Macaire gave a more general impression:

> As one penetrates the interior, the trees become more numerous, larger, and some kilometres from rivers, one comes to hardwood species, such as mahogany, rosewoods and others. But one must penetrate to about 20 km from the coast or the borders of the river to find a part not exploited, containing trees of 1.2–1.3 metres diameter and 45–50 metres tall.
>
> (Macaire 1900: 34)

Table 2.2 Early estimates of the area of the forest zone and area of high forest in Côte d'Ivoire (millions of hectares)

	Chevalier 1909c	*Gros 1910*	*Sargos 1928*	*Mangin 1924*	*Meniaud 1922*	*Meniaud 1933*	*Aubréville 1937*
Forest zone	12	12	11–12		12	11	11
Forest cover	6	7.2–7.8	7.5–8	7.5		8	Refuses to estimate

A little later, in 1938 (but still well pre-dating the mid-century onset of devastation as represented today), Aubréville noted that 'in travelling outside the major routes, one can no longer hold any illusion. Entire regions are covered only in secondary forest.' He was precise about the location of 'primary' forest: 'There exist still some large blocks of primary forest in uninhabited or uneasily settled regions, in middle Comoé and between Sassandra and Cavally, for example, but in all the inhabited country, the secondary forest dominates' (Aubréville 1932: 239). Later he specified that by secondary forest he meant fallow vegetation destined to be felled when judged sufficiently mature, suggesting that much of it had been quite recently farmed (Aubréville 1937: 238–9, 1947a). It seems conceivable that the CTFT study misinterpreted this mosaic of relatively young bush fallow and tree crop farms as forest.

Population and forest cover

Consideration of the relationships between population and land use also provides cause to question assertions of 11–12 mha of high forest cover in 1955, or indeed of a vegetationally intact forest zone at the turn of the century. In the regions west of the Bandama river, rice is a major crop. Here Chevalier calculated that population densities of only about 7 people per km^2 were sufficient to cultivate the entire territory within preferred fallow lengths of c. 12–15 years: 'This is no exaggeration, as around villages of c. 200 people, the forest is destroyed in a radius of between 5–7 km' (Chevalier 1909c: 47–8).[3] In 1955 at least three regions west of the Bandama had significantly greater densities: Daloa (7.2/km^2), Man (10.6/km^2) and Gagnoa (16.4/km^2). Only the regions in the extreme south-west (Sassandra 2.2/km and Tabou 1.6/km) had smaller populations, in keeping with descriptions of these areas as the country's major forest preserves (Republic of Côte d'Ivoire 1988). The accuracy of forest cover estimates derived from population and land use calculations is clearly open to some question, but it cannot be disputed that the regions were peopled and farmed at the turn of the century, contrary to the image suggested in modern statistics.

In the east of the country, too, many areas had significant populations in the early to mid-twentieth century. In the region of Moronou, for example, the administrator Marchand documented 62 villages and as many farm-hamlets in 1907 in an area of around 350,000 ha (Ekanza 1981). If each village farmed a radius of only 4 km, this would have accounted for the entire land area. Probably some areas would have been more and others less inhabited, and the vegetation thus a mosaic of forests of different ages, merging with kola, palm, plantain and other plantations. But it would be completely incorrect to consider this region to have been 'forest'. Similar village and population densities were found throughout the Aoussoukrou and Ouellés regions of the Nzi Comoé; a glance at early maps of the whole

Cercle of Nzi-Comoé suggests equally large numbers of villages throughout an area of 1.6 mha, all within the forest zone. The only area in this Cercle where early French administrators might have found lower populations was that devastated and depopulated by Samori Touré's army in 1895. Only east of the Comoé river do certain regions appear to have been less populous. In the south-east, for example, Perrot (1976, 1982) suggests a figure of c. 16,000 people in Ndenye, an area of c. 600,000 ha; i.e. around 2.4 per ha. She argues that the region had never been more populous, and indeed that such populations had increased from a quarter of these levels in the previous two centuries. Despite some regions of low population, levels of farming population at the turn of the twentieth century across most of the Ivorian forest zone were certainly incompatible with an image of extensive pristine forest at this time.

Several further aspects of settlement and population dynamics had a significant influence on forest cover.

Roadside relocation

As Aubréville pointed out in 1937, village movement and relocation were common phenomena in Côte d'Ivoire, adding to the impoverishment of forest left behind the impoverishment of the new forest land moved into. Together with the extensive shifting cultivation, he wrote, such causes 'explain how the area of secondary forests can be large in Côte d'Ivoire, despite the low population density' (Aubréville 1937: 240).

Late colonial authors pointed to considerable deforestation near roadsides. Both in 1937, and when revisiting Côte d'Ivoire in 1957, Aubréville stressed how much forest along the roads had been felled. Mangenot noted that even in less populous regions, patches of deforested land 'nevertheless coalesce, above all along roads, in bands which, each year, become bigger, growing, damaging the forest a little deeper' (1955: 21).

A number of recent authors have associated this pattern of settlement along roadsides with the arrival of immigrant populations, roads providing the avenue in. Lanly, for instance, argues that forest disappearance is partially governed by road development, with cropping extending outwards from existing roads and inwards into the forest along new tracks (see Monnier 1981: 190; Sayer *et al.* 1992). Yet while this pattern of development around roads may indeed be indicative of agricultural expansion into the forest today, it is certainly not a new phenomenon, nor one linked exclusively – or even principally – to migration into the forest zone.

Many existing villages and hamlets were in fact forced to move early in the colonial period under a policy of relocation and consolidation of inhabitants to the new roadsides. It could be argued that at this time, the pattern of forest denudation along roadsides reflected more such forced resettlement than an increase in felling by new settlers. Forced relocation of villages was intended to improve administrative and political control over the population and to locate workforces in a way convenient to the maintenance of the roads. As Mundt notes in a basic text on Côte d'Ivoire: 'The population remains linearly distributed along the main roads as a result of such resettlements, which continued into the 1930s in southwest Ivory Coast' (see Mundt 1987: 8; see also Schwartz 1971: 22). As Avenard *et al.* emphasise for the Seguela–Vavoua region, present population distribution reflects forced movement during the 1920s and 1930s, not the recent settlement of migrants along roadsides (Avenard *et al.* 1974: 53, see also Lena 1979 for the south-west of the country).

While resettlement may have precipitated new forest felling near roadsides, it may also have abandoned equally large areas away from the roads. Having previously been farmed, these acquired the possibility of returning to high forest. This is the case, for example, in parts of the south-west between the Cavally and Sassandra rivers, where recent immigration has been into forest regrowth lands which were forcibly depopulated under the orders of Captain Schiffer between 1907 and 1910. Much of the Ivorian forest area away from roads may similarly represent regeneration following abandonment c. 1920s. This implies that actual observations of deforestation between 1910 and 1950 cannot necessarily be taken to indicate an increase in the area farmed and of forest lost; they may reflect a mere relocation of farming and a change in forest quality but not area. Aubréville noted this:

> Since our establishment in the country, the population has again moved. The first maps drawn of the colony by officers and administrators carry the names of villages which one searches for in vain today. Nothing signals them now but the secondary forest, and sometimes, when recently abandoned, some cotton trees planted by former inhabitants.
>
> (Aubréville 1937: 239–40)

Depopulation

While much of the forest zone may have been fallow or recent secondary forest following relocation, a large part was also secondary following overall depopulation. Several regions of Côte d'Ivoire were massively depopulated during colonial wars, most notably against the Tepos in 1899–1901, the Baoulé in 1902–1912, the Dan in 1905–8, the Beté in 1906, the Guro in 1907 and the Dida in 1909. The Dan, Beté, Guro and Dida inhabit a huge region of the Ivorian forest zone. Regions considered by today's forest statisticians to have been more or less pristine 'forest' at the turn of the century, presumably without people or history, were well able to fight protracted wars against the French.

The effect of these wars on the Baoulé people is the best documented and can serve as an example. Weiskel (1980) compares successive population estimates to provide an idea of the magnitude of the devastation involved. Referring to observations during the period of Baoulé prosperity from 1895 to 1899, Delafosse (1908) estimated the total population in the Baoulé region to be nearly 2,000,000, or at minimum, 1,300,000. Albert Nebout placed the number in Baoulé territory at this time as slightly lower – 1,000,000 – but Captain le Magnen suggested a higher figure of 1,500,000. Even if one takes the most conservative of these estimates as representative of the pre-1900 population, it becomes apparent that the military activity had a marked impact on the Baoulé region. Clozel (in Weiskel 1980) estimated that by 1902 the Baoulé population was only 642,548. By 1911 the administrator of the northern Baoulé district, formerly the most populous of all the Baoulé regions, estimated that the population level stood at 130,585. If one generously doubles this figure to represent the additional population of the southern Baoulé region, the total still reaches only c. 260,000 in 1911. In reality this figure may be an over-estimate, for the population of the Baoulé region by 1916 was placed at only 225,000. In short, from 1899 to 1911 the Baoulé population suffered an enormous decline, whether from direct deaths in battle, ensuing famine and disease or outmigration towards other regions of the colony or into the neighbouring Gold Coast (Weiskel 1980: 208–9).

The significance of this depopulation becomes clear when compared with official population figures for the entirety of Côte d'Ivoire in 1921, which stood at c. 1.5 m, rising to only 2.5 m in 1955. In short, Baoulé populations alone around 1900 were not far different from the entire Ivorian population in 1921, and only just less than the Ivorian population in 1955. While Weiskel suggests that dispersal accounted for much of the population loss, this begins to look untenable in the face of such huge figures of countrywide decline, and indeed evidence of decline elsewhere. In the west, Viard, for example, noted that 15,000 people fled from the Cercle of Guiglo (Taï forest region) at the time of French occupation, equal to about half of the remaining population in 1932 (at c. 32,000) (Viard 1934a, b, c). In the extreme south-west, populations did not increase between 1906 and 1972 (Massing 1980). Indeed, it seems strongly possible that populations in this region and inland of it were higher in the 1830s than they were from 1920 until the 1970s. When Reverend Wilson visited the Ivorian town of Grabo, 60 km inland in 1836 (a source overlooked by Ivorian historians), he found a populous region with a thriving rice-farming economy. The economy and inland trade to inhabitants of the interior forest region were so dynamic, based on the cocoa and coffee introduced prior to French occupation, that Tepos forces were able to sustain a war against the French (Wilson 1836; see Wondji 1963).

In many parts of the country, the demographic shocks from military and social dislocation in the early colonial period were compounded by illnesses such as the 1918–19 flu epidemic. The area farmed and cleared would have been further reduced by labour shortages engendered by conscription, whether of soldiers for the 1914–18 war, or for porterage and road building. Depopulation for a variety of reasons, then, allowed forest regrowth in its wake.

Intriguingly, as Mangenot (1955) notes, the vast uninhabited forests between the Cavally and Sassandra rivers also show evidence of earlier occupation. In these apparently most 'virgin' of Ivorian forests, foresters find artefacts of settlement including iron furnaces, pottery, tools and charcoal in soil pits dug to reveal ecological variables (e.g. Sayer *et al.* 1992: 134; van Rompaey 1993). Findings of this nature suggest that the past inhabitants of these forests were farmers. At least a third of the Taï National Park is recognisably secondary, or 'new' forest (Bousquet 1977 in Lena 1979: 73), although systematic archaeological exploration in this region has yet to be pioneered.

The area of the forest zone

A further question emerges concerning the area of Côte d'Ivoire's 'forest zone'. By 1909, Chevalier had asserted that:

> The limits of the forest of Côte d'Ivoire are today known. It is in reality less vast than one had originally thought. We think that it measures about 12,000,000 ha. This forest is followed by a very wooded savanna, presenting wide gallery forests along water courses. . . . We believe that it is not exaggerated to consider half of the surface of the supposed virgin forest as occupied by this forest of recent formation, much less rich in wood.
>
> (Chevalier 1909c: 35)

According to Chevalier's definitions, the 'forest of recent formation' was forest either maintained within fallow cycles (10–15 years growth), or which had recently been abandoned from management. It had a poor flora, with only about 30 tree species, compared with – he suggested – 350–400 in the primary forest (Chevalier 1909c).

Figures for the area of the forest zone varied considerably around Chevalier's estimate until the late 1950s, ranging from around 11 to 13 mha. However, assessments made since independence have generally suggested that the forest zone is very much larger, from the 15.7 mha given by Lanly (1969) and Guillaumet and Adjanohoun (1971), to the extraordinary 21 mha in Sayer *et al.* (1992).

How can this discrepancy be accounted for? There seem to be several possibilities. The first comes from a cynical viewpoint which recognises that recent international observers have had little inhibition for exaggeration. It gives their own work more powerful rhetorical effect, whether for their own cause or career, and there is little accountability. A figure of 21 mha of 'original' forest, for instance, almost doubles the extent of past forest at the stroke of a keyboard, at the same time raising the extent of deforestation to dramatic levels which demand to be noticed.

Second and more importantly, at least 2.3 mha of this discrepancy between early and present estimates can be explained when one realises that certain areas on the northern margins of the forest zone which today are considered to lie within it were classified as outside by earlier observers. These include (a) areas south and south-west of Bondoukou; (b) areas north of and west of Groumania; (c) north and east of Daloa; and (d) south and west of Seguela (see Map 2.1). Equally, several early authors considered a strip of land 20 km inland from the coast (c. 1 mha) as outside the forest zone, as it lacked forest, whether this was because of the presence of lagoons, littoral savannas, mangroves or heavily settled farmlands.

It appears that early descriptions concerning the extent of the forest zone referred to precisely that: where the zone carrying moist semi-deciduous forests ended. In contrast, it seems that the past extent of the forest zone in modern works has come to be defined climatically as the zone in which it is supposed that forests could survive and, so it is presumed, did once exist. These figures thus include areas which were savannas in 1900 or at some other date, which today's observers nevertheless think to have been derived from forest. Not that theories of savanna derivation in a climatic forest zone are new. This same deduction lent support to early analysts' views that the forest boundary was in retreat under local land use (Chevalier 1909c;[4] Aubréville 1938); an opinion often still heard (e.g. Gornitz and NASA 1985; Fair 1992; van Rompaey 1993).

The regions of Vavoua and Man, in particular, provide intriguing insights into representations of the forest zone. Lanly (1969) considered Vavoua to lie well within the forest zone. Several analyses today frame farming and other changes in this region in terms of massive deforestation. Hervouet and Laveissière (1987), for example, relate deforestation there to disease epidemiology; they assume that the Vavoua region carried forest in the 1950s which has subsequently been destroyed. But this schematic vision of deforestation is not supported by any comparative analysis of air photographs.

Yet from all early maps (Chevalier 1912a; Meniaud 1922; Mangin 1924), it is clear that Vavoua was classified outside the forest zone in early years of the twentieth century. An indication of the area's vegetation can be gleaned from the account of Thomann, the first French officer to arrive there in 1903. On the road from Balogué to Bologué, some 25 km south of Vavoua, he described the first clearings: 'One crosses a first very large clearing, with immense roniers standing here and there at long intervals, as in Baoulé. At last the mission has left the dense forest' (Thomann 1902–3: 632). Between Bologué (Borogué) and Gaibi (5–10 km south of Vavoua) there was thicket and 'brousse épaisse' (without clearings) but whatever this was – whether he means tall grass savannas, or relatively wooded land – it was not forest. Thence he proceeded to Vavoua, where the vegetation began to 'open up seriously' (presumably into grasslands). If the region around Vavoua was in the forest around 1950, it was certainly not so in 1900.

In 1938, maps put Man on the edge of the forest zone, or outside it. Today's vegetation maps by contrast tend to locate Man within the forest zone. Aubréville, defining the forest zone so as to exclude the area, was nevertheless convinced that forest could exist there, and so once had existed:

> In a recent epoch, Man was situated in the forest. The state of the mountains which immediately surround this small town prove it incontestably. They are still today occupied by an extremely degraded forest, invaded by high grasses. The taller trees which remain, protected by rocks still grouped in ravines, are an absolute proof.
>
> (Aubréville 1938: 206)

He went on to cite the prevalence of several species indicative of retreating forests, including *Trema guineensis*, *Albizzia zygia* and *Ficus mucoso* (Aubréville 1938). By the road from Man to Seguela, Aubréville noted forest thicket for 30 km. High 'primary' forest existed only where there were sacred groves (1938: 210). He deduced that these groves were relics and used them to identify 'what used to cover these mountains'. Near Seguela, he noted that the villages were all installed in the interior of forest islands, on the summits of small hills and dominated by the major species of deciduous forests: *Antiaris africana*, *Celtis zenkeri*, *Ceiba pentandra*, *Triplochiton scleroxylon*, *Khaya grandifoliola*, *Blighia sapida*, *Cola cordifolia*, etc. He observed similar village forest islands between Man and Touba. Yet in many, Aubréville claimed that all the stages in their progressive destruction by farming could be seen. In one village, he observed:

> [T]he disparition of the forest which once clothed it is made evident by a circle of *Anogeissus schimperi* which has the village as its centre. To the exterior of this circular curtain of trees extends the wooded savanna; to its interior the savanna of *Pennisetum purpureum* with some cotton trees and *Ficus* around the village.
>
> (Aubréville 1938: 206)

More generally, Aubréville argued that:

> When one observes attentively this transition zone between the continuous forest and the savanna, one comes to identify clearly the areas which, until a recent date, were covered by forests. Some indicators suffice, such as two or three silk cotton or *Cola cordifolia* above a patch of high grass. These two species seem to be the most resistant of the large trees to fire.
>
> (Aubréville 1938: 81–82)

Yet as we have noted in Chapter 1, evidence from the nearby Kissidougou region of Guinea suggests that it is incorrect to assume that forest islands such as these, either with villages at their centre or associated with abandoned village sites, are the relics of past, more extensive forest. While this may be the case, in Kissidougou, where this was also assumed (and also by Aubréville 1938), the forest patches turned out to have been established in savannas. The landscape was half full of forest islands, not half empty. Moreover, forest islands in what Aubréville considered to be the later stages of decay can be those in the early stages of creation. From the descriptions of the forests, and the species cited, these forest patches appear to be very similar to those in Kissidougou. As Aubréville himself notes, *Cola cordifolia* is 'distributed' by inhabitants who eat the sugary fruit and appreciate its shade. *Ceiba* is used in fencing. And both trees can take their forest form with tall, straight boles without growing in forest. Air photographs of these forest patches also show their similarity to those found in Kissidougou. Monnier (1981: plate 128), for example, presents an aerial view of a forest island near Semien (near the Sassandra river, between Man and Seguela): a 'curious white patch of a village nestling at the heart of a crown of trees'. There is every reason to suspect that the forest patches which Aubréville observed had been created by people.

Is the area of the forest zone increasing?

A third reason why estimations of the extent of the forest zone have increased is simply that the area of forest or forest fallows (which, according to older classification methods, defines the forest zone in opposition to annually burnt grasslands) has actually been increasing since 1900, not contracting. In short, a case can be made that much of the northern part of the Ivorian forest zone has recently been derived from savanna. The evidence for encroachment of the forest zone on savanna comes from several sources: from studies of vegetation change which use historical and present data; from oral history in today's savanna, transition and forest zones; and from archaeological evidence.

Several accounts suggest that areas now well within the forest zone have been savanna in the recent historical past (Map 2.2). In a forest area south-east

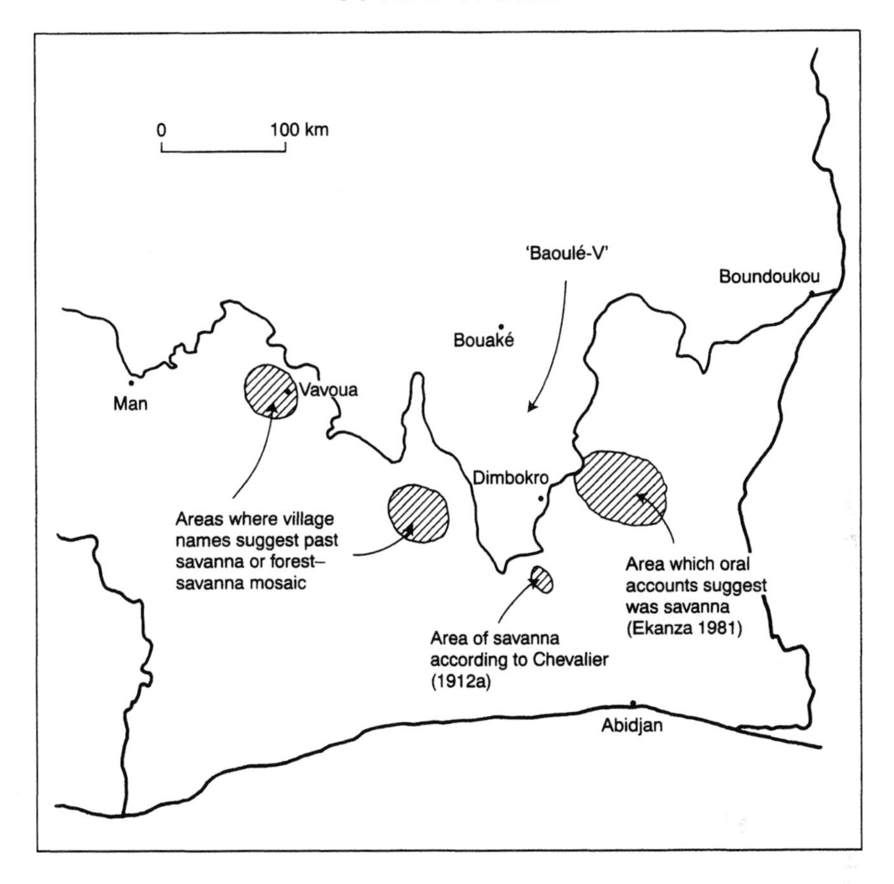

Map 2.2 Areas in today's Ivorian 'forest zone' which historical evidence suggests to have been savanna in recent centuries.

of the 'Baoulé-V', Ekanza, for example, was perplexed to hear from elders in the Moronou region that:

> [T]he Agni traditions of origin make reference to the savanna as being the form of vegetation which dominated Moronou at the period of settlement. Village sites during this period of invasion were chosen in function of the openness of savanna. Brobo,[5] for which the origin goes back to the earliest time of Agni settlement in Moronou, was built on a site in savanna.
>
> (Ekanza 1981: 59)

In a taped interview in 1981, the 85-year-old Eonan Messou suggested that 'Nanan Sangban, head of the Essandane, abandoned Brobo, where game was getting rare, for the preferential site of Bongouanou, which offered at once savanna, water and animals . . . The foundation of Arrah and other villages obeyed the same imperatives.' From here [Bongouanou] to Arrah, tradition from the period of settlement reports a game-filled savanna extending as far as the eye could see (1981: 59–60). The author Ekanza finds it difficult to uphold such an affirmation, which implies that in the space of only two centuries, from the eighteenth century when Agni arrived to 1907, the forest expansion has been so rapid that it would have absorbed in its progression all of the savanna. Yet in neighbouring Guinea we found the encroachment of forest into savanna in the transition zone to have covered tens of kilometres in only three decades (Fairhead and Leach 1996a).

West of the Baoulé-V, in Guro country again in what is now the forest zone, Deluz gives the meanings for several Gouro village names which indicate the presence of savanna when they were founded. For example, just east of Sinfra the village name of Koumodjé means 'forest island in the middle of savanna', and Dianambroufla means 'man who installed the village in elephant grass' (1970: 175). Diedenoufla in the same region means 'buffalo, son of elephant grass', also signifying an area of savanna. Deragon (15 km north of Vavoua) literally means 'open savanna'. Frefredou (40 km east of Daloa) refers to the place as open 'as if locusts had eaten all the grass'. These are all villages which are considered by Lanly (1969) to be within the 'forest zone'. As Blanc-Pamard and Peltre (1987) argue, village names are not an entirely reliable guide to prevailing vegetation since they might highlight outstanding rather than generalised features. Yet these names do, at least, suggest the presence of a mosaic of vegetation, rather than the solid block of trees presumed by many of today's analysts.

Such a vegetation history of forest advance is entirely consistent with the findings of several ecological studies made in Côte d'Ivoire since the 1960s. In the savannas of the Baoulé-V, for example, villagers themselves suggest that 'where one cultivates, the forest advances', and research on forest dynamics at the forest–savanna boundary shows just that (Spichiger and Blanc-Pamard 1973: 199). This finding was supported by those of Adjanohoun, who writes:

> When one asks Baoulé elders about the origin of savannas, they affirm that their ancestors, 200 years ago, found the same vegetation formation which they call *kakie*, meaning wooded savanna. The old cultivators recognise that they have contributed to the degradation of islands of dense forest, but they make you observe that when they fallow their ex-forest fields, exhausted by cropping, the latter regrow not in *kakie*, but in forest fallow. They affirm equally to remember the existence of *kakie*, once found within the dense forest, and which has today disappeared, entirely invaded by this [forest]. For them, the grassy savannas are natural, and there is not an actual phenomenon of savannisation. On the

contrary, it is the forest which gains on the savanna, and this despite their action.

(Adjanohoun 1964: 131)

Mitja (1990) and Mitja and Puig (1991) show how in the humid savannas north of Man, when savanna land on slopes (constituting most of the land area) is farmed under shifting cultivation, there is astonishing regeneration: 'Not only do inhabitants not degrade the region, rather they improve it' (1991: 390). Sites initially covered with low, woody savanna, come to acquire, after 7 years of farming and a long fallow of 10–40 years, vegetation with a more closed, denser tree cover and fertile soils favouring infiltration, having more surface holes from worm and termite activity.

A further study shows forest cover increasing over savanna (Spichiger and Lassailly 1981) between the 1950s and 1970s in the Beomi region. These trends continue in the Baoulé-V savannas: 'Lamto savannas are characterised by . . . their dynamic evolution towards forest' (Menaut and Cesar 1979: 1197). 'In 20 years, tree density in annually burnt plots has increased by c. 30 per cent' (Dauget and Menaut, unpublished data cited in Menaut *et al.* 1991: 136). The latter suggest that this might be due to an intensification of 'wave-like' or cyclical climatic patterns, and that extreme events or episodic, concurrent disturbances should then be responsible for the maintenance of savannas in the very long term. But this overlooks the evidence above, that at least in the areas south-east of Lamto the savannas have not been maintained.

The reasons for forest expansion are thought by some to be climatic, at least in part, owing to rehumidification. Aubréville, who pioneered the idea of forest recession and savannisation under human management – a view which still dominates policy circles – came to change his mind over this issue, and to pioneer a very different theory. He suggests that 'the climatic conditions which permitted the establishment of savannas were those of a relatively recent period, and . . . we are seeing again today the development of forest colonisation following climatic rehumidification' (Aubréville 1962: 30). He noted how coastal savannas have ceded to forest under coffee and guava cultivation. Just how much of Côte d'Ivoire's forest zone has been savanna in historical times remains to be ascertained. The evidence presented here suggests that this is an urgent climatic, ecological and anthropological research question. It hardly needs stating that the possibility removes once and for all the validity of the assumption that under climates where forest can exist today, it once did exist – a false assumption that we will have cause to refer to again and again in later chapters. Clearly too, this analysis contradicts those assessments which considered the forests once to have extended further north. Yet despite Aubréville's reversal, the theory of savannisation still captures the imagination of many modern conservationist writers and policy-makers.

Conclusion

It would be imprudent and impudent of us to suggest that the question of forest cover change in Côte d'Ivoire is now resolved. Nevertheless, the little evidence reviewed here suggests that current orthodoxy concerning forest cover change in the country requires serious revision, something that will necessitate more focused analysis. If one avoids presuming the extent of past cover but instead uses historical sources and time series data, the rate of change in Ivorian forest cover looks very different. The orthodox estimates of forest loss in literature published between 1960 and 1990 suggest that Côte d'Ivoire had between 14.5 and 16 m ha of forest in 1900, began to lose it dramatically around 1955, and by 1990 had only about 2.7 m ha remaining. Thus estimates of forest loss between 1950 and 1990 stand at around 330,000 ha/year. Yet c. 7–8 m ha would seem to be a generous figure for forest cover c. 1900, less than half of the orthodox view. Côte d'Ivoire may have had similar forest cover in 1955. Subsequent rates of forest loss remain quite high, at c. 130,000 ha/year on average, but this is only about 40 per cent of generally accepted loss rates (see Table 2.3).

Equally, attention to historical sources concerning past populations and land use suggests that interpretations of forest cover change in terms of one-way loss have obscured dynamics of very different kinds. Forest growth or regrowth on depopulated farmland, and the expansion of forest on its northern margins, encouraged both by certain farming practices and by climatic rehumidification, both seem to have been highly significant processes in Côte d'Ivoire's forest history. As we will see, these are themes which recur in subsequent chapters.

Table 2.3 Forest decline reconsidered in Côte d'Ivoire (millions of hectares)

Date	*Forest area (FAO/UNEP 1980)*	*Reconsideration*
c. 1900	14.5	c. 7–8
end 1955	11.8	c. 7–8.8
end 1965	9.0	c. 6.3
end 1973	6.2	c. 5.5
end 1980	4.0	–
end 1990	2.7	2.7
Total	11.8	c. 5.3–7.3 m ha

Note

If we accept a smaller definition of the forest zone, and consider the Lanly (1969) finding that what was defined as forest declined from 75 per cent to 53 per cent of the area, a drop in forest cover from perhaps 9 m ha in 1955 to 6.5 m ha in 1966 could be considered. This remains on the upper end of contemporary and believable estimates for 1955, and lower end for 1965.

Notes

1 It might be thought, as Guillaumet and Adjanohoun note in passing in their report, that they had tried to update their analysis of land use from the 1955 photographs before publishing in 1966. But this seems doubtful given that they did not use the 1966 air photographs, and that ground truthing to verify land use change across the whole of Côte d'Ivoire could not possibly have been effected to a sufficient level of precision in the time available. When finalising their report, they did not have access to Lanly's use of 1966 air photographs and work.

2 This is also mentioned in his 1909 report on Ivorian forests: 'Towards km 120 in the region of Tranou there appears savanna in forest.'

3 Mangenot similarly suggested that a ring of farm and recent fallow land with a radius of 5–10 km around villages was normal (Mangenot 1955).

4 'Undoubtedly, the African virgin forest extended far further than its present limits' (Chevalier 1909c: 43).

5 This village is given in some traditions as being the first agglomeration of Moronou, prior to the foundation of Ehuikro, considered by other sources as the first settlement.

3

LIBERIA

The vast bulk of Liberia is considered to lie within the forest zone, with areas of savanna restricted to a coastal strip and to the extreme north-west of the country; savannas which most authors consider to be anthropogenic in origin (Voorhoeve 1979: 19; D'Azevedo 1962a: 530). Dorm-Adzobu exemplifies a dominant present-day opinion that: 'within living memory, about 90% of the land surface of Liberia was covered with mature forests, composed entirely of broadleaf or hardwood tree species' (1985: 3). Not only has deforestation over the last century been seen as a steady progression, but this has been forecast to continue. Thus the US Environmental Profile of Liberia (USAID 1980) estimated that the continuing demands of shifting cultivation and logging would lead to the complete disappearance of the country's remaining primary forests within another 10 years.

Like the case of Côte d'Ivoire, the Liberian case presents an opportunity for critical examination of forest cover statistics. But more than for Côte d'Ivoire, we pursue here the dialogue and dissonance between forestry perspectives on forest history and social and demographic analysis of the country's past.

A picture of Liberian deforestation as unilineal is supported by a number of social anthropological and historical works. As we will show, these forward seemingly powerful arguments concerning population movements and growth, and agricultural transformation, to account for the gradual loss of Liberia's 'original' high forest cover from around 1500, albeit accelerating in more recent times. Hasselmann sums up this argument:

> Numerous data show that before the historical shifts of the West African population in the sixteenth century, Liberia was almost totally covered in high forest. With the arrival of the coastal peoples, then the migrations from the Sahel towards the north of Liberia, large areas have been transformed into anthropic savannas.
>
> (Hasselmann 1991: 51)

However influential within the social sciences, as we shall examine, these views of the longer-term history of Liberia's forests actually rest at odds with the analyses current among the most influential foresters. The latter hold that most of Liberia's forests consist of old secondary regrowth following the decline of farming populations which 300 years ago were probably larger than today's. This chapter examines Liberia's forest history in the light of these contradictory positions. Drawing on historical sources hitherto omitted from these debates and on the works of modern historians, we find the historically inflected forestry position more tenable. Indeed, we would attribute more significance to this than do foresters, who seem to minimise its importance and significance for conservation.

Critical analysis of longer-term history is also necessary to contextualise assertions made about forest cover change during the present century. The

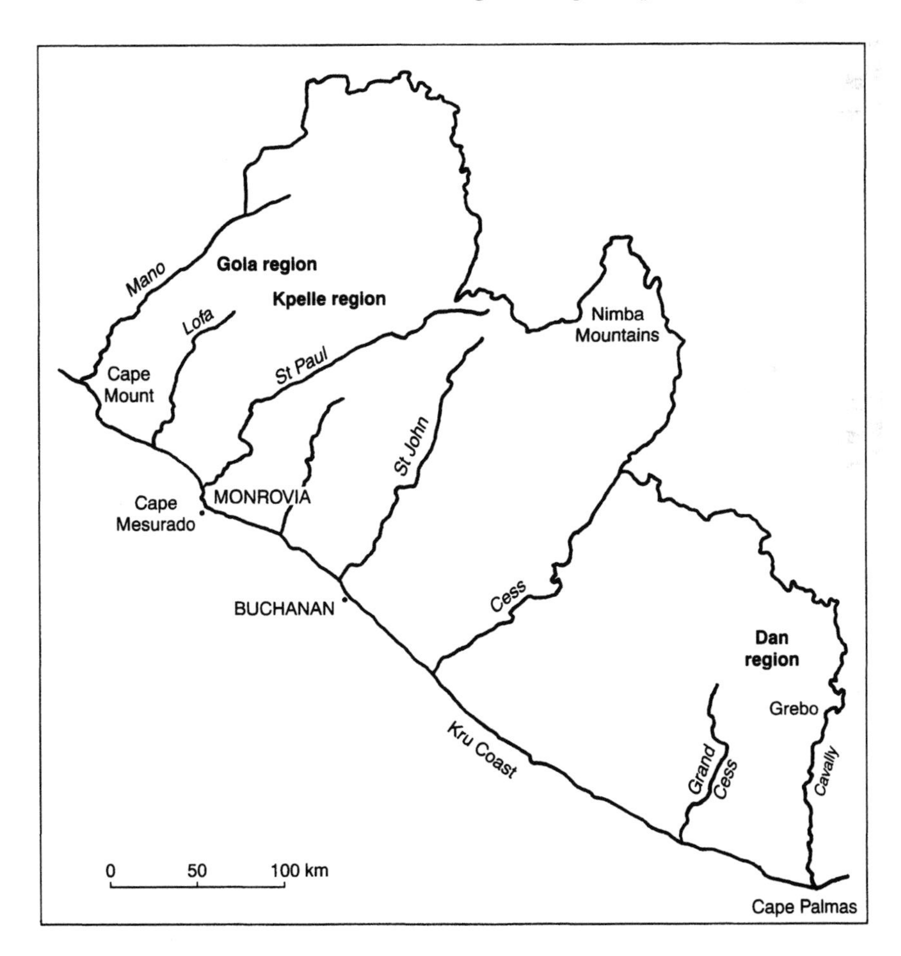

Map 3.1 Regions of Liberia described in the text.

last part of the chapter examines a range of figures purporting to describe the extent and rate of recent forest loss, exposes the contradictions between them, and concludes that the most influential analyses have much exaggerated twentieth-century forest loss in Liberia.

Deforestation and the settlement frontier

A view of Liberian history in terms of a settlement frontier pushing progressively into sparsely inhabited forests has, until recently, framed social and historical analysis of the region. It has been articulated most influentially by D'Azevedo (1962a), whose analysis informed the subsequent work of a generation of anthropologists (e.g. Murphy and Bledsoe 1987; Nyerges 1988; Davies and Richards 1991; Leach 1994). D'Azevedo argues that northern and western Liberia 'seems to have been a relatively unexploited rain-forest area, inhabited by a sparse population of hunters, fishers, and gatherers practicing minimal cultivation, into which both Sudanic and European influences intruded at about the same historical period: that is, between the 15th and 17th centuries' (1962a: 524).

Between the seventeenth and nineteenth centuries, a frontier of Sudanic grain crops (rice and fonio) and farming techniques, and associated population increase, is purported to have moved from north-west to south-east. D'Azevedo linked these transformations to the southwards movement of Mande-speaking peoples who at once forced 'indigenous peoples' further into the coastal forests and were responsible for introducing slash-and-burn techniques and their iron tools. 'The process of transformation of rain forest into secondary growth and grassland, which had begun two or three centuries earlier to the north of Sierra Leone, was now well underway in the Central West Atlantic Region' (1962a: 535).

D'Azevedo argues that 'these developments had affected all of Sierra Leone by the end of the sixteenth century, but the influences in north-western Liberia were as yet minimal' (1962a: 534–5). Thus, he states elsewhere:

> Much of the coastal territory which the Gola now occupy was a dense uninhabited rainforest until as late as the eighteenth century. Their slow westward migration from the interior region of Komgbaa in the mountains of north-eastern Liberia began in the seventeenth century under pressure from the powerful savanna empires of the western Sudan. Like many of their neighbours, the Gola retreated into the coastal forests where they became widely dispersed through migration, warfare and the slave trade.
> (D'Azevedo 1962b: 12–13)

That Gola are relatively recent rice farmers, he states, is indicated by their farming and dietary practices and their oral traditions. He argues that south-eastern Liberia remained relatively isolated from these influences until the late nineteenth and early twentieth centuries. Indeed, 'the forests of south-eastern and central Liberia appear to represent an area of hiatus between the spread of Sudanic agricultural patterns from the north east, and the root and wild forest crop complex from the easterly tropical forests' (1962a: 520–21); the latter being the intensive yam cultivation practised by Baoulé peoples in Côte d'Ivoire, west of the Bandama river. The Krahn, Kru and Grebo peoples living in the forest hiatus area, he suggests, 'have not been intensive farmers until quite recently'. 'Minimal yam and cassava cultivation was a mere supplement to fishing and forest gathering' (1962: 521), with associated low population densities and small village sizes (50–90 huts) until the twentieth century. More intensive farming, and particularly rice cultivation, 'can be dated no earlier than one century ago for many of the southern Bassa, Kru and Grebo peoples. Among the Grebo, rice cultivation has appeared only in the present century through the influence of Liberian settlers and government programmes' (1962a: 521). He extends this logic across the border into Côte d'Ivoire, where he argues that the Guru did not plant rice until this century and that the Gagu have not yet adopted it.

In short, D'Azevedo's analysis for Liberia conforms with analyses popular among many colonial observers of neighbouring countries: the gradual southwards shift of Sudanian populations, technology and ecology. It supports a view of Liberia as little-disturbed forest in the north-west until as late as the eighteenth century, and in the south-east as late as the early twentieth century.

In a sophisticated analysis, Ford (1992) has recently provided an alternative model for the deforestation frontier for north-eastern Liberia, centred on the links between kola production and the movement of settlements. The area currently inhabited by the Dan is characterised by large numbers of abandoned village sites rich in kola trees, whether planted by earlier inhabitants and preserved, or planted more recently over abandoned villages. Ford argues that Dan farmers have long planted kola trees around their settlements, for their economic and social value. When a settlement site is abandoned, the house and kitchen garden sites would also be planted over, becoming in effect a kola orchard on the fertile soils of the settlement site. He argues that kola orchards thus multiply as village sites shift.

He goes on to suggest that because, for the Dan, the presence of kola is associated with land that is out of bounds for farming, kola in effect 'spoils the bush' for shifting cultivation. In order to maintain fallow cycles for shifting cultivation, farmers are therefore forced to bring new land into the fallow cycle. He thus advances the hypothesis that 'kola cultivation promoted Dan forest expansion by placing land suitable for food crops out of production, thereby shortening the swidden cycle and "quickening" forest penetration' (1992: 51). This longstanding process was significantly hastened in the late nineteenth and early twentieth centuries as production expanded to meet the growing kola trade encouraged by Manding merchants.

Ford therefore argues the case for forest penetration, but not simply as a wave of shifting cultivators into virgin forest; rather, a more complex dynamic reflects the ways that farmers integrated kola production and abandoned settlements with their bush fallow cycles. He also provides a powerful critique of the commonly held view that people such as the Dan were 'pushed' into the coastal forests by the southwards movement of Manding peoples, showing instead how an autonomous expansionist dynamic within the forests was linked to the Dan's own economic development, and that it was this which served to 'pull' Manding traders.

Nevertheless, his argument concerning the presence of old village sites and the forest penetration dynamic which they enhance is made in the context of an assumption of continual population expansion or at least stasis. While the reasons he cites for settlement abandonment include disease, warfare and village consolidation, he implies that populations simply move. He does not consider the possibility that the existence of so many abandoned village sites might indicate significant population decline.

Forest history and the forest frontier

This dominant vision of progressive forest conversion, whether led by Sudanic crops or the kola economy, does not sit easily with historical evidence suggesting that large areas of Liberia have experienced periods of population decline, sometimes dramatic, and associated forest regrowth.

Ironically, it is the analysis of a forester cited frequently by the modern conservation lobby which most clearly contradicts the portrayal of recent and unilineal deforestation. During his assessment of Liberia's forest cover in the 1940s, Karl Mayer walked 2,300 miles (3,700 km) through the Liberian forests both to ground-truth air photographic analysis (in which he had been a wartime expert) and to assess the national population. On the basis of his observations of forest composition and abandoned village sites, all of which he was required to note, dating their abandonment, he came to very different conclusions regarding forest history.

Mayer argued that 'there are many indications that the population has been greatly reduced in the past two or three decades. Greatest population losses are believed to have occurred recently in the Eastern Province, where abandoned villages are, in some sections, very common' (1951: 16). He argued that:

> The parts of Liberia which today show no signs of occupancy during recent centuries are few and scattered. . . . Aside from mountainous areas, which did not encourage occupancy and clearing, the most

> primitive high forests were found in a belt lying between 30 and 50 miles inland across the eastern Province, from the River Cess to the D'Bor River drainage. It is probable that this area, some 1,500 square miles [c. 400,000 ha], constitutes a core of the original forests which was never disturbed and is quite representative of the ancient forest cover.
>
> (Mayer 1951: 25)

Otherwise, his impression over sizeable areas of Liberia was that it was an 'over-used, worn-out country of great antiquity' (Mayer 1951: 25). In this context, he suggested:

> It is possible that as recently as 300 years ago [i.e. c. 1650] there was considerably less high forest area in Liberia than is found today. The subsequent sharp decline in population, as heavy tribal warfare and slaving activities exerted their drain, possibly coupled with the ravages of new diseases, would have permitted many cleared areas to complete their reversion to high forest. . . . Only by such an hypothesis can the writer explain the over-all compositional inferiority of Liberian high forests as compared with those considered typical of the Ivory Coast and the Gold Coast.
>
> (Mayer 1951: 25)

His interpretation of depopulation and forest regrowth – continuing well into the twentieth century in some places – was thus quite the opposite of orthodox social analysis which has viewed populations as expanding over the last few centuries, with the growth accelerating in the twentieth century.

Importantly, Mayer was not just deducing depopulation and forest regrowth from botanical form, but linking vegetation analysis with systematic observations and inquiries concerning past settlements. Nevertheless, most of his botanical deductions concerning the secondary status of the forests were confirmed in the national forest inventory of the 1960s, prepared with German assistance (Sachtler 1968), and by the forester Voorhoeve (1979). Voorhoeve cites and accepts Mayer's view that 300 years ago Liberia may have been much more densely populated than at present. He gives as botanical evidence 'the occurrence of extensive single dominant forests . . . the occurrence of forests where the secondary character becomes evident from the species composition, and the presence of relics of human occupation such as graves, ancient roads etc.' (1979: 18). Of the high forest area, which he estimated at c. 35 per cent of the country, he considered 'untouched' forest to be extremely rare, finding it only in the gorges of the Nimba mountains.

While Voorhoeve's botanical deductions concerning single species dominance clearly provide an instance of reading vegetation history from present form – a method which we have criticised – he is nevertheless careful to distinguish different types of single dominant forest, some of which appear to give no historical information. Thus, when a single species dominates a particular forest storey, 'a certain even-aged character can be suspected, and the forest is probably of secondary origin: an incidental abundant and successful regeneration during the low bush or young secondary forest stage resulted in a mature single-dominant forest' (1979: 22). Such forests were particularly dominated by either *Parinari excelsa*, *Monopetalaanthus compactus* or *Gilbertiodendron preussii*. He distinguishes these forests, reflecting past farming, from others which probably reflect edaphic conditions. For example, forests dominated in all storeys by *Tetraberlinia tubmaniana* are probably edaphic, associated with sandy soils. The forester Holsoe earlier drew the same conclusions about *Parinari excelsa* dominated forests, noting the tendency for farmers to leave this hard-wooded and useful rough-skinned plum standing in fields after other trees have been cut (Holsoe 1961: 11; see also FDA/IDA 1985: 8).

Mayer's deduction concerning the history of Liberian forests was incorporated into forestry canon and gained credibility through support from established botanical methodology. His estimation that 300 years ago there was less forest than at present was repeated by Holsoe and Voorhoeve, who are cited in the recent assessments of FAO (1981), the World Resources Institute (Repetto 1988: 82) and the World Conservation Monitoring Centre (Sayer *et al.* 1992). Van Rompaey (1993: 53) notes: 'even single-dominant forests were said to have originated after farming' (see also Hart 1990). Nevertheless, this historical context is treated as of only peripheral interest in these modern reports; useful for interpreting the broad nature of forest composition, yet in a restricted way which pays no attention to the specific legacies of past settlement and land management. Equally, this history is commonly portrayed as only a brief hiatus in a more important longer-term history of loss of 'original' forest. Most significantly, it has become accepted to consider this as an episode 'about 300 years ago' (e.g. Repetto 1988), distancing it and rendering it barely relevant to present populations and land claims, despite Mayer's observations of depopulation well into the twentieth century.

Critique from social historical sources

These forestry analyses, which challenge views of Liberia's social and demographic history in terms of a settlement frontier, gain further support from more recent historical research which critiques the frontier theory, and from a number of nineteenth-century accounts of the Liberian interior.

The clearest critique has come from the historian Adam Jones and his interpretation of the accounts of a Dutch trader in the 1630s (presumed to be the source of the Kquoja account in Dapper 1668; see Hair 1974). This account described in some detail the economy and society of the large

Kquoja kingdom whose rulers resided near Cape Mount on Liberia's extreme western seaboard. Using the superb Kquoja account and other minor seventeenth-century sources, Jones argues that D'Azevedo's frontier analysis 'underestimates the amount of agriculture which was being practiced, often independent of any obvious "Sudanic" influence' (1983a: 26). These sources described large-scale rice cultivation not only near the coast, but inland far to the north and east where 'most rice' was found. They gave no evidence to suggest that rice farming was anything but well established among these non-Mande-speaking peoples. They also described an extensive inland trade in rice seed, cloth and slaves. In short, 'if we look at the whole range of food, the image of the people of this area as merely hunters, fishermen and gatherers recedes into the very distant past. Neither Europeans nor Mande-speaking Africans brought about the kind of transformation that has sometimes been suggested' (Jones 1983a: 27).

We should note that the Kquoja source did mention the existence of 'a great wood of 8 or 10 days' journey in length', near the site of today's Gola forest, straddling the Sierra Leone–Liberia border (Dapper 1668). In the account this forest is delimited by the countries of the Gala-Vy and Hondo peoples to its east and north-east, and the 'right Gala' on its other side (presumably in today's Sierra Leone). The same source suggested that the then inhabitants of 'Gala-Vy' originated in peoples driven out of their Gala country by the people of Hondo, who sought new habitations among the Vy. One can speculate that this forest wilderness of delimited extent, buffering the Kquoja kingdom from peoples to the north-west, might itself have been of recent origin, again following depopulation. The wars which caused the Gala (Gola) to be refugees at this time were probably those associated with the infamous Mane conquest of this region, and appear to have been noted briefly by Manuel Alvares 1615: chap. 13).

While Jones' critique refers largely to west and north-west Liberia, Massing's examination of early sources from south-eastern Liberia led him to be equally critical of the frontier thesis. Massing (1980) points to documentation of rice trading from the Kru coast in 1624 and suggests that rice may have been produced in this region since well before this. He notes that European visitors mention purchases of rice at points along Liberia's eastern seaboard such as from Rio Cestos (e.g. de Fontheneau 1554: Villaut 1669).

While we have found no historical information regarding the area of northern Liberia covered by Ford's analysis, this region borders on the Dan-speaking regions immediately across the border in Côte d'Ivoire. Here, Schwartz (1971) shows the zone to be relatively densely populated in present times and, on the basis of oral history, argues that it was an important pre-colonial trade corridor encouraging high populations around intense economic activity, dating back at least as far as the sixteenth century (the period of the Kquoja state). He states that 'If the actual inhabitants did not arrive until the beginning of the eighteenth century, oral tradition suggests a much older occupation' (Schwartz 1971: 15). Traces of earlier populations in the Dan region, in particular in the vicinity of Toulepleu, exist in the form of stone ruins, including the remains of village walls within which one finds pottery in what is today the thickest of forest (Schnell 1949; Schwartz 1971). While oral tradition appears to be silent on the origins of such villages, Schwartz suggests that they might have been built by earlier populations (Schwartz 1971: 38).

A larger number of historical sources document conditions in the nineteenth century. Sources inland in south-east Liberia from the 1830s to 1860s, for example, suggest the presence of long-established rice-farming populations, giving no indication that rice was a recent adoption. Massing cites Hall (1836), who in a tour to the interior up the Cavalla river in the extreme east of Liberia described the banks as 'teeming with its rich waving harvest of rice' (1980: 119). A year later, in the same region, Wilson was 'not a little surprised both at the extent of cultivation and the quality of rice. Portions of the rice through which our path lay had attained its full growth and quite as good as any that I had ever seen in the rice country of South Carolina or Georgia' (1836: 193, cited in Massing 1980). Inland, in regions D'Azevedo considers to have adopted rice only in the present century, Wilson (1836: 387) observed rice cultivation in many villages. That rice was a long-established and ritualised crop is indicated in the notes of Alexander Crummell, the celebrated American pan-Africanist, who paddled far up the Cavally river in 1860. 'It was the rice-gathering season. The great labour of the year was over, and the people were preparing themselves for the "Dance of Joy" for the harvest. The women were adorned with their best, and all the people full of delight and song'[1] (see also Hoffman 1862).

The Reverend Dr Savage made a journey inland from Cape Palmas in 1839:

> The rice farms are very extensive, and, at one time, are seen, as we ascend the river (through a small opening among the trees made for a landing place) expanding far beyond, into fields of many acres; at another, the 'bush' being cleared away to the very verge of the river, unfolds to the eye an immense expanse, waving in all the luxuriance of nature.
>
> (Savage 1839: 156)

On the Cavally river he 'frequently met with canoes, laden with palm oil, rice, and cassadas, destined for market at Cape Palmas' (1839: 157). These accounts also describe the presence of much larger settlements than the small hamlets which D'Azevedo associated with this area. For instance, Wilson described the village of Saurekeh as having a population of 1,500 to 2,000, and Kay a population of 500, 'walled around with split timber'

(Wilson 1836: 243). These villages lay in the vicinity of the coast or on riversides and, given the importance of rivers for transport at this time, it would be unwise to generalise from them. Walking tours made into the far interior in the 1860s also suggest the presence of large numbers of sizeable towns. The missionary Brownell, for example, journeyed from Cape Palmas to Mount Gede (Brownell 1869), passing twenty-seven towns during a walk of around a hundred miles. While he described neither agriculture nor town sizes, in many towns he preached to congregations of several hundred people, in one case 400. Again, 'pathside biases' might limit the generalisability of these estimates. But none of these accounts, whether referring to agriculture or population, seem to bear any relation to D'Azevedo's characterisation of eastern Liberia at this time. As Massing sums up:

> [T]he presence of rice cultivation at the Kru Coast is well documented for the periods of first contact with the Kru Coast and there is also evidence that surplus production supplied slave traders, Americo-Liberian colonists in Monrovia and legitimate traders in the eighteenth and nineteenth century. . . . Furthermore, sales of food commodities other than rice seem to have been standard practice in coastal villages since the early seventeenth century.
>
> (Massing 1980: 119)

Accounts of inland journeys on foot throughout central, western and northwestern Liberia along assorted routes between 1830 and 1870 also speak of large populations of predominantly rice farmers in these regions and give no indication whatsoever of rice being a recently adopted crop. The Americo-Liberian missionary George Seymour lived for about two years around 80 miles (130 km) east-north-east of Bexley in Pessey (Kpelle) country. Neither on the way to this station through Bassa-speaking country (Seymour 1856) nor at this station (Seymour 1858) did he give any description to suggest that the rice farming was novel. Quite to the contrary, he described how 'This section of the country is cleared to a great extent for farming purposes, rice being the principal product, of which there appears a good supply of the very best quality' (1858). This picture is confirmed in the detailed travel accounts of Sims (1860), Blyden (1869) and Anderson (1870, 1874) in what are today the Gola, Kpelle, Belle and Loma regions of Liberia.

While all of these travel accounts, in both eastern and western Liberia, described parts of their journey as through 'heavy forest', this appeared within an overall description as patches in an otherwise heavily farmed landscape. It could also be argued that the pathsides or riversides on which these descriptions focused may have carried vegetation untypical of the rest of the landscape. Nevertheless, the travellers also attempted to climb hills and mountains to get better views, and such descriptions are invariably of populated farm and fallow land with patches of forest. From one summit near the Cavally river, Wilson, for example, wrote that 'the compass of vision in any direction could not have been less than thirty miles. We saw three native settlements and my men pointed out high trees which denoted the site of several others' (1836: 241). From a hilltop in Loma country, Seymour described: 'The summit and sides are mostly planted in rice . . . we counted one hundred and forty farms . . . the city of Solong, with five considerable towns in the south-east. The palm-tree makes a large part of the remaining forest, and can be counted by thousands . . . the city of Vogemer and three towns are visible to the north-east' (Seymour 1860).

Despite these descriptions of large and vibrant farming populations, a number of these authors also made reference to prevailing warfare and depopulated regions; Sims, especially, described such conditions in Pessey (Kpelle) country in the mid-nineteenth century. They generally accredited abandoned villages not to the social dynamics of settlement movement (as does Ford), but to warfare, enslavement and depopulation. Indeed, the coastal slave trade continued in Liberia well into the mid-nineteenth century and the inland slave trade to the Mande and Fouta regions further north continued into the early twentieth century.

Massing (1980) describes the decline of the population and economy of the Kru coast associated with the political changes under incorporation into the Liberian state. Initial armed revolts in 1910 of most interior peoples against the Liberians were put down ruthlessly by the Liberian frontier force under American command, destroying most of the settlements north of the Cess river (1980: 102). With World War I, following economic collapse an uprising on the lower Kru coast was again suppressed, destroying most settlements from Sinoe to Sasstown. Massing argues that local officials used the Frontier Force as a means to raid among the unarmed Kru peoples for captives who were later sold as labourers to Spanish or Portuguese colonies. Such practices seem to have continued until the late 1920s, with sporadic uprisings against Liberian authority continuing until the 1930s. As the archives attest, after the wars came 'deserted villages, neglected farms, hungry times' (1930 report cited in Massing 1980: 103). As Massing argues, many of those who did not suffer forced emigration fled to Sierra Leone, Ghana and Nigeria. Holas (1952: 142) also described many deserted towns in eastern Liberia.

Elsewhere in the country, Akpan (1982–3) has described how uprisings took place in Kpele and Bandi country (1911–14), Kissi country (1913), among the Gio and Mano (1913–18), among the Gbolobo Grebo (1916–18), among the Gola and Bande (1918–19), among the Joquele Kpele (1916–20), and at Sikon in 1921. Again, these uprisings were ruthlessly suppressed by the Liberian state, often with intentional depopulation of the areas concerned (see Akpan 1982–3). As noted by the Harvard expedition to Liberia in 1927 (Strong 1930), exactive conditions similar to those in the

Kru Coast Area – whether hut taxes or forced labour for roads, plantations and porterage – were enforced on populations throughout much of the interior.

In short, historical evidence provides some support to Mayer's observations of major depopulation during the late nineteenth and early twentieth centuries. Evidence suggests that while the dating of '300 years ago' by foresters was indeed a time when Liberia's populations were greater and forest cover less than today, this situation clearly continued into the nineteenth century. Whereas species dominance in various strata of the forest may attest to earlier periods of depopulation, it is the abandoned sites observed earlier this century which attest to more recent events; in areas which may, over the intervening decades, have come to bear 'old growth' secondary forest. In short, in many areas of Liberia forest regrowth may be rather more recent than most foresters are assuming.

In this context, it could be argued that the limited farming of rice in the early and mid-twentieth century portrayed by D'Azevedo, especially in south-eastern regions, did not reflect the slow and gradually accelerating adoption of rice as agricultural practices transformed. Rather it might have reflected a decline in rice cultivation following periods of warfare, depopulation and subsequent labour scarcity. Shortages of male labour for land clearance were further accentuated by conditions of forced labour for the state. Furthermore, between 1910 and 1930 at least, the requisitioning of foodstuffs by Liberian soldiers and travelling state representatives was notorious. The Harvard expedition, for example, noted that after the periodic visits of state representatives 'the towns and villages are frequently left at least in temporary destitution, for apparently almost everything is taken away. Goats, poultry and other food supplies . . . and sometimes even the young or more attractive girls disappear' (Strong 1930: 113–14). Under such conditions, rice production might well have been further discouraged in favour of less resource-demanding root crops that, stored in the ground, were not so amenable to expropriation. Such transformations in farming under structurally similar conditions have been noted to favour cassava over grains in former Zaïre (Fairhead 1989).

Forest cover change in the twentieth century

The historical evidence discussed above provides an important context for considering the various estimates of twentieth-century forest cover change in Liberia. Clearly, assumptions that virtually all of Liberia was covered in its original or mature forest around the turn of the century cannot be upheld. Yet this was suggested by Dorm-Adzobu (1985) and by Parren and de Graaf, the latter considering the original moist forest area to be 7.3 m ha in around 1900 (Parren and de Graaf 1995: 28).

A number of analyses have used assorted statistics to portray precipitous forest loss during the twentieth century. Gornitz and NASA (1985) compare a forest cover of 6.5 m ha in 1920 with 2 m ha in 1985, to suggest a loss of 4.5 m ha over the 65-year period; a rate of 69,000 ha/year. These figures are based on Zon and Sparhawk (1923) and FAO (1981). Parren and de Graaf are hardly less dramatic, comparing 7.3 m ha with a present cover of 3.9 m ha, derived from 1985 air photographs to suggest a loss of 3.4 m ha, or 40,000 ha/year over 85 years.

These analyses over the century timescale are complemented by others which suggest precipitous loss at a more modern decadal timescale. These are summed up in Sayer *et al.* (1992):

> FAO estimated the annual rate of deforestation for the years 1981–5 to be 46,000 ha. . . . The FDA/IDA (1985) figures together with unpublished estimates from the author J. Mayers, give an annual rate of deforestation of 94,600 ha for the years 1983–9, which is an annual loss of about 2%. This figure accords well with the decline in amount of forest from the 1979–82 analysis of FDA/IDA (1985) to that shown on the 1987 satellite images.
>
> (Sayer *et al.* 1992: 217)

These assertions of deforestation on both the century and the decadal timescale can be questioned, if not falsified. As we now go on to show, they depend on the highly selective use of statistics and definitions of forest, ignoring or minimising counter-evidence.

The first systematic measurement of Liberia's forest cover was that made by Mayer in the 1940s. As an expert interpreter of aerial photographs, he used US air force air photographs taken in 1945–6 covering practically the whole country and spent two years ground truthing the images. For 1946, he estimated that Liberia's high forest covered only c. 3.5 m ha, or 38 per cent of the total land area. This high forest included both completely undisturbed forest and old secondary regrowth (see Holsoe 1961). A further 20 per cent (1.9 m ha) of the land area Mayer characterised as 'broken bush', which he interpreted as forest with a broken canopy which originated in partial cuttings for agricultural purposes. Characterising Mayer's broken bush within the category forest (as Holsoe 1961 did) would give a total cover of 5.4 m ha, although this is a very loose definition of forest. It may be that broken bush referred to a lower type of vegetation than that usually treated as 'high forest' by today's analysts, perhaps of relatively recent regrowth of forest escaping from the farm–fallow cycle, following the depopulation which Mayer had noted. The rest of the country was covered by low bush or non-forest land.

In 1960–67, a new forest inventory was carried out by the Bureau of Forest and Wildlife Conservation in co-operation with the German forestry mission (Sachtler 1968). This produced a total forest cover of 2.5 m ha.

Several of the analyses of precipitous decline cited above rely on this figure or its derivates. Perhaps the most influential of these derivates was the FAO 1980 Tropical Forest Resource Assessment for Liberia, which updated the 1968 figure according to a population growth–deforestation model, to suggest a total 1980 forest area of 2 m ha, and a projected area of 1.8 m ha in 1985 (FAO 1981). Gornitz and NASA used the FAO figure for 1980 in asserting precipitous forest loss during the twentieth century. Notably the FAO was aware of Mayer's work, but preferred to use the German assessment as it was more recent.

But as Hammermaster (FDA/IDA 1985) argues, the German inventory was deeply flawed as a basis for assessing forest cover. Intended primarily as a timber inventory, it did not cover all of Liberia but only major forest areas, including those declared as national forests, and it recorded only forests occurring in large contiguous blocks. Furthermore, Hammermaster shows that the German forest mission hugely overestimated the total area of Liberia – as 11.1 m ha instead of 9.6 m ha. This meant that it hugely underestimated percentage forest cover. When the FAO 1980 report presented the country's percentage forest cover as 19 per cent, it compounded these two errors: an underestimate of 1960s forest cover and an overestimate of Liberia's area.

In 1985 Hammermaster used 1979 and 1982 air photographs to assess forest cover. High forest was found to cover 4.8 m ha, or 50 per cent of a total land area of 9.6 m ha (FDA/IDA 1985: 17). This survey thus revealed the huge error in the FAO 1980 assessment for Liberia, forcing a dramatic reassessment of Liberian forest cover change. Hammermaster's survey defined 'high forest' as primary or old secondary in nature, with a closed or almost closed canopy exceeding 30 m in height. Notably, and in contrast with the German assessment, the survey covered the whole country and all forest areas down to a size of 40 ha.

More recently, several studies have estimated present forest cover on the basis of satellite data (1 km NOAA AVHRR). Using 1987 images, Paivinen and Witt (1988) estimated a forest area of 4.26 m ha. Using 1989 and 1990 images, Stibig and Baltaxe (1993) again estimated Liberian forest cover, this time at between 4.55 and 4.75 m ha. In the latter, poor image quality and lack of ground truthing prevented a clear distinction being made between closed and 'disturbed' forest; the figure is thus for total forest area (i.e. neither fields nor bush fallow), broadly comparable with Mayer (1951), although perhaps less so with FDA/IDA (1985) and Paivinen and Witt (1988), which were both more rigorous in their distinctions between forest types. Paivinen and Witt's figure referred only to closed forest with a canopy cover of more than 40 per cent, leading to the expected difference in results. Thus, recent satellite data only serve to confirm Hammermaster's dramatic reappraisal of the FAO (1981) assessment.

Comparing these recent assessments with Mayer's forces a dramatic reappraisal of Liberian forest cover change in recent decades. It would appear that between 1947, when Mayer estimated closed canopy and broken bush to cover 5.4 m ha, and 1990, when Stibig and Baltaxe found c. 4.65 m ha of closed and disturbed canopy forest, there was a loss of only c. 0.75 m ha. As Mayer suggested, the figure for 1947 was probably higher than earlier in the century, in retrospect forming what may be considered as a high point in Liberian forest cover over the last few centuries. A loss of 0.75 m ha is only about a fifth of the 3.4 m ha loss since 1900 suggested by Parren and de Graaf, or 16 per cent of the 4.6 m ha loss since 1920 suggested by Gornitz and NASA (1985). Even using the stricter definition of forest in Paivinen and Witt, forest loss since 1947 would be 1.1 m ha, or about a quarter of recent assessments. Contrasting estimates of Liberian forest cover at different times are summarised in Table 3.1 and Map 3.2.

While recent air and satellite data should have forced a re-evaluation of recent rates of forest loss, this appears not to have been the case. Indeed, despite recent data, modern estimations seem to be claiming higher rates of loss than ever before. For instance, while Sayer *et al.* (1992) accept the FDA/IDA (1985) reassessment of forest cover area, and its critique of the FAO (1981) assessment, they surprisingly cite the FAO figure for the rate of deforestation, taking this to be 46,000 ha/year over 1980–85. FAO's calculation was based on a model of supposed deforestation linked to

Table 3.1 Forest cover estimates for Liberia

Date	*Area (ha)*	*Reference*
1900	7,300,000	Parren and de Graaf (1995)
1920	6,475,000	Zon and Sparhawk (1923)
1950	5,520,000	Haden-Guest *et al.* (1956)
1951	5,684,800	Mayer (1951)
1953	8,900,000	FAO (1953)
1968	2,500,000	Persson (1974)
c. 1972	2,500,000	Myers (1980)
1975	4,856,232	Frank and Gorgla (1975)
1980	2,000,000	FAO (1981)
1982	4,790,000	FDA/IDA (1985)
1983–4	4,800,000* (2.5 m ha 'broken')	FDA/IDA (1985)
1985	1,770,000	FAO (1981; estimate from population model)
1990	4,633,000	FAO (1993)

Note

* Of this area, national forests (government reserves) accounted for 1,699,681 ha (6.5 per cent of which is under shifting cultivation, 2.4 per cent is exploited by concessions, 26.1 per cent is exploited by non-concessions and threatened by shifting cultivation, and 65 per cent is non-exploited forests), and other forests constituted 3,100,319 ha.

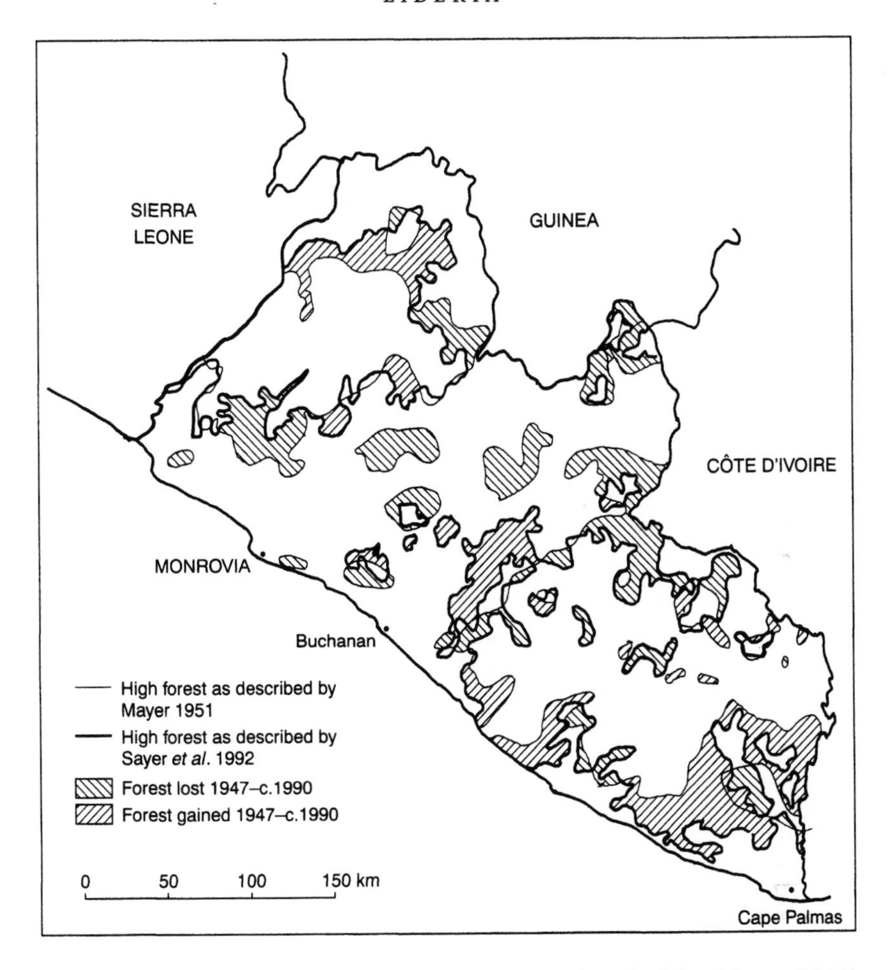

Map 3.2 Principal high forest areas of Liberia as described by Mayer (1951) compared with areas of rainforest according to Sayer *et al.* (1992).

increased agricultural land clearance under prevailing population growth rates and was in keeping with the FAO's exaggerated figure for the extent of forest loss over recent decades. It is hard to reconcile such a loss rate with the subsequent re-evaluation of forest cover. This in turn calls into question the validity of the population–deforestation model – for a revised loss of 0.75 m ha between 1947 and 1990 suggests a rate of loss of only 17,400 ha/year. More surprisingly still, Sayer *et al.* (1992) consider the FAO's forest loss rates as a conservative estimate. By comparing FDA/IDA (1985) figures with unpublished figures from J. Mayers for 1989, they calculate an annual forest loss nearly twice as large: 95,000 ha/year for 1983–9. The nature and source of Mayers' figures are not detailed. Nevertheless, they suggest that the calculation is also in keeping with a comparison of FDA/IDA (1985) with Paivinen and Witt (1988). However, these high figures can be reconciled with older data only by supposing that deforestation massively accelerated in the 1980s. Since the same authors suggest that the principal cause of deforestation (95 per cent of all forest clearance) has been smallholder agriculture, this supposition would require evidence of a major 1980s episode of rural population increase, which has not occurred (see Hasselmann 1986). One might note that Liberia's rural population in mid-1990 was c. 1.5 million, an increase of only 400,000 from 1947 (see Mayer 1951; Sayer *et al.* 1992).

To maintain views of precipitous deforestation, authors have at times also misrepresented the data in past studies, whether intentionally or not. Thus, for example, Mayer's category of 'broken bush' (1951), as Holsoe (1961) makes clear, refers to land partially cleared for agriculture. Yet Parren and de Graaf (1995) ignore this agricultural inference, claiming that it had merely been logged over; a claim which supports the view of forest little disturbed by farmers earlier in this century, but which is internally contradicted elsewhere in their own work where they state that 'in Liberia the export of timber was almost non-existent until the 1960s' (1995: 32). Data questioning high rates of forest decline have been provided by Hasselmann (1986), who showed that FAO (1981) estimates of the rate of clearance were twice those shown by his surveys. This scale of deforestation is borne out in his data given for the rate of deforestation calculated in selected regions of Liberia over 30 years.

Conclusion

Liberian deforestation during the twentieth century may represent only 20–30 per cent of the area suggested by most authors concerned with deforestation. Furthermore, it seems that the early twentieth-century period taken by many international analysts as the baseline for rapid loss of 'original' forest cover may actually have been a high point following earlier depopulation and forest regrowth. It is quite possible that forest area in Liberia was increasing during the first half of the present century. This possibility has become apparent in this chapter through exploring the dissonance between different lines of argument within forest science on the one hand, and social and demographic history on the other. Historical analysis of vegetation change in strands of forestry science, as well as the critical views of demographic and agrarian pasts in the recent work of historians, firmly question established images of population, land use and deforestation in terms of an inexorable one-way settlement frontier.

The extent to which northern parts of Liberia's forest 'regrowth' following depopulation have effectively involved the colonisation of forest in regions which had not been forest prior to this occupation has never been considered. Forest history at the northern edge has been regarded only in terms of unilineal encroachment and savannisation – with the pitfalls referred to earlier when considering the neighbouring Ivorian forest. That this latter question needs to be addressed in Liberia is clear from historical work on the Ziama reserve in Guinea's forest region, just to the north of the Liberian border. Here, long-inhabited and populous savanna areas in the 1850s are now semi-deciduous rainforest reserves following complete depopulation c. 1870 to 1910 (Fairhead and Leach 1994a). Under these historical conditions encroachment into forest reserves and uninhabited forests may merely be the work of populations reclaiming their ancestral farmlands, which had earlier been alienated from them when the reserves were established. As subsequent chapters will show, there is similar, albeit disparate, evidence for such processes across the West African forest belt.

Notes

1 Extract of a letter by 'Alexander Crummell, B.A. Queens' Coll. Camb.', an ordained 'Negro' clergyman of the American Church, now teaching at Cape Palmas – to John Kitton, October 1859, forwarded by Kitton to the Royal Geographical Society. Archives of the Royal Geographical Society, London.

4

GHANA

Introduction[1]

This chapter aims to appraise the many analyses of the nature, rate and extent of forest loss which have recently been made for Ghana. As for Côte d'Ivoire and Liberia, we assess such analyses in the light of evidence from historical sources. In this critique we demonstrate not only how prone are modern statistics to exaggeration, but how such tendencies also prevailed among early colonial forest statistics and representations of forest quality. This chapter is, however, able to elaborate further the relationships between past populations and present vegetation quality, underlining the significance of periods of depopulation as well as population growth in West Africa's landscape history.

Numerous studies assert that a large part of Ghana (around 8.6 m ha) belongs to the 'closed forest zone'[2] where climate and soil conditions enable forest to be supported (Map 4.1). They assert that most of this zone was covered in closed forest in the 1890s, but that with the exception of around 1.8 m ha of reserves, this forest has since been lost due to encroachment by farmers and loggers. Furthermore, some areas which were formerly part of the forest zone are now characterised by savannas.

In a recent major study of Ghana's forest composition and botany, Hawthorne and Musah (1995) describe eloquently the orthodox view concerning the demise of the country's forests to their remarkable present pattern:

> When the Forestry department officers and workers started to draw out their planned shapes for the forest reserves in the 1920s and 1930s, with the boundary lines running often as rather arbitrary straight paths between concrete pillars, enclosed on all sides by forest, they could hardly have imagined that within 60 years the pattern of their handiwork would be visible from orbiting satellites. Although it was against such an outcome that the boundaries were cut, it is highly remarkable that satellite pictures

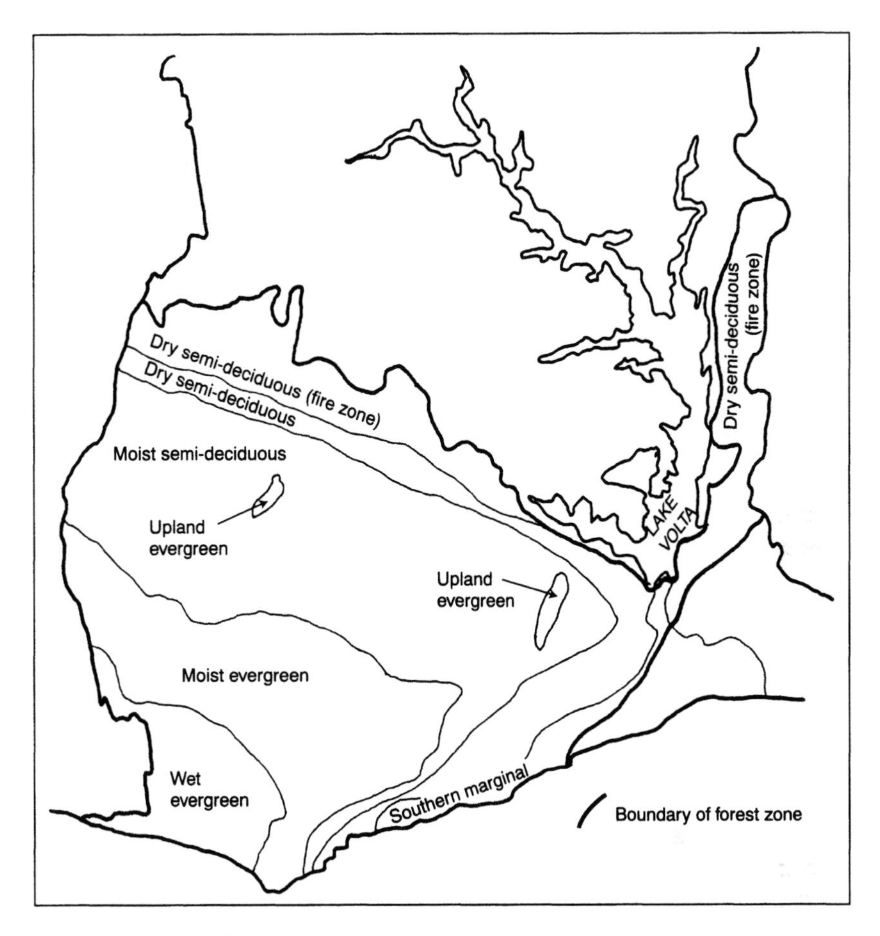

Map 4.1 Ghana's forest zone and its component vegetation types as accepted in modern analyses (Hall and Swaine 1981).

> of the forest remaining today coincide with the plans of the thirties, and that there is virtually no significant area of forest outside the reserves. . . . The wave of deforestation has only recently arrived en masse at the gates of the reserved estate, and it is now that maximum pressure exists for the walls to come tumbling down.
>
> (Hawthorne and Musah 1995: 11)

These authors go on to note that only about half of the nation's 1.8 m ha of forest reserves is 'in reasonable condition', with about 14 per cent of the reserves having 'no significant forest left' (Hawthorne and Musah 1995; see Map 4.2).

Whether through repetition or research, Ghana has acquired such a

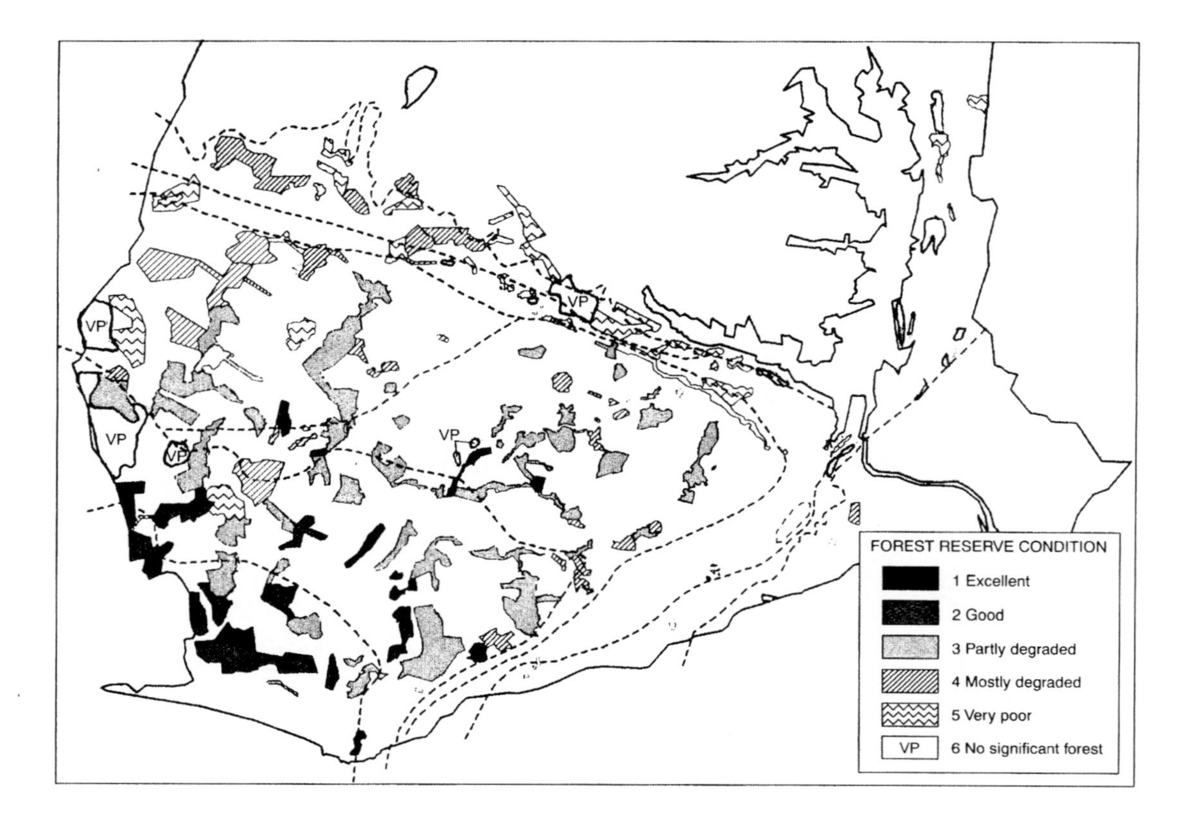

Map 4.2 Ghanian forest reserves and their condition (Hawthorne and Musah 1995: 17).

reputation for massive and relentless forest loss and degradation in the international literature that it now frames much historical, geographical and social anthropological analysis of the country. Foresters working in Ghana are generally well aware of the more complicated historical past of their forest zone and its people, especially in their own particular regions of operation, not least because they must engage with the political legacy of the forest's human history in their daily work. Indeed, Hawthorne and Musah caution analysts against an overly romantic view of Ghana's forest past, noting that many forests had once been farmed, most had been inhabited, and many had burnt from time to time. But the idea that the Ghanaian forest zone was recently pristine forest is nevertheless a beguiling first approximation, and it still orientates the way most analysts first approach forest issues. No one can deny that there has been extensive forest loss in Ghana during the present century. We will argue, however, that the extent of twentieth-century deforestation is being hugely exaggerated in international scientific circles, and with it the extent of the problems supposed to have resulted. This is distorting not only present social and historical analysis within Ghana, but also climatic modelling of importance to the region's future.

Despite the numerous modern analyses of forest cover made using air-photographic and satellite imagery, no studies have compared old and new photographs in assessing vegetation change in Ghana in a way that would shed light on these questions. Equally, virtually none of the studies making claims about vegetation history has been rigorous in compiling historical data (documentary or oral) concerning past vegetation, against which to compare more recent conditions. This chapter is a first step towards examining assertions made about Ghana's forest history in relation to some of the historical data available for this richly documented country, with the aim of shifting the terms of debate and encouraging further research. We address several principal questions: Is it valid to assume, as many authors do, that the extent of the forest around 1880–1900 was the extent of the 'forest zone' as it is now defined? Did all of what is today considered as the forest zone carry forest in historical times? How was the forest zone used in the late nineteenth century and before? In particular we want to explore how much of the land under forest in 1890 had once been farmland but reverted to forest due to depopulation and, relatedly, how many of the national forest reserves now of poor quality were that way when they were reserved. Finally, we address the dynamics on the forest zone's northern and eastern margins: has forest been retreating at the expense of savanna in historical times as most work assumes?

Ghana's forest cover at the turn of the century

Romantic estimations of past forest cover and of subsequent deforestation dominate in the forest conservation and policy literature, suggesting that

Ghana's 'forest zone' was more or less intact forest in the 1880s. Most assessments suggest a forest cover figure of 8–10 m ha around 1900, now reduced to less than 2 m ha.[3] These are the figures which appear in the ODA Forest Inventory Project Seminar Proceedings, for instance, where Frimpong-Mensah asserts:

> [At] the turn of the century, Ghana had over 8,800,000 ha of forests. . . . Only 4,200,000 ha of this remained by about 1950. The estimate for 1980 puts the forest area at about 1,900,000 ha. This means that Ghana has lost over 75% of its tropical forest within this century, due to inefficient agricultural practices (shifting cultivation) and over-exploitation.
>
> (Frimpong-Mensah 1989: 72)

Ghartey, also working for this major forest inventory, asserts that around 1900 Ghana had 8.1 m ha of forest, compared with 2.1 m ha today (Ghartey 1990, in Parren and de Graaf 1995). Fair (1992) states that Ghana's rainforests have been reduced from 8.2 million to 2 million since 1900. Ebregt (1995: 6) notes that 'at the turn of the century, it was estimated that Ghana had 8.8 million ha of forest. By 1950, this had fallen to 4.2 million and by 1980 it was estimated at 1.9 million'. Gornitz and NASA (1985) suggest a decline from 9.9 m ha in 1920 to less than half of this by the 1950s and a decline from 4.8 m ha in 1938 to 1.7 m in 1980. These works are cited by others. For example, Ghartey's figure is cited by Parren and de Graaf (1995); contributors to Sayer *et al.* (1992) cite Frimpong-Mensah; and Mather (1990) cites Gornitz and NASA. In keeping with these figures, the World Bank (1988) has estimated that closed forest has been lost at an annual rate of 75,000 ha since the turn of this century. As most authors point out, if deforestation has recently declined, this is simply because little forest remains and what remains is reserved.

The closed forest zone from which forest is thought to have been lost is usually delimited as shown in Map 4.1. The areas supposedly occupied 'at origin' by the various forest types distinguished within it are given in Table 4.1.

Those who consider Ghana to have been deforested during the twentieth century view the country's social and economic history in broadly corresponding terms. This historical perspective is expressed by Hall (1987: 33), who argues that large-scale deforestation began about a hundred years ago when 'the abolition of the slave trade had led to a decrease in warfare between states, increasing population and the greater development of forest products such as palm oil, kola and copal' (citing Dickson 1971). Virtually all authors characterise deforestation as having occurred as a frontier, or a wave, as new areas of forest are progressively lost to farming.

However, the determination of past forest cover in these estimates is

Table 4.1 Area occupied at origin by components of the closed forest zone in Ghana

Vegetation type	*Area (ha)*
Wet evergreen	657,000
Moist evergreen	1,777,000
Moist semi-deciduous	3,289,000
Dry semi-deciduous	2,144,100
Upland evergreen	29,200
Southern marginal	236,000
South-east outliers	2,000
Total	8,134,300

Source: Sayer *et al.* (1992).

highly questionable. While some authors have simply assumed that the entire forest zone was forest, even those using early forest cover estimates have done so uncritically. Gornitz, for example, asserts that his figure for 1920 forest cover is based on Zon and Sparhawk's (1923) estimations.[4] Yet a reading of Zon and Sparhawk does not support this. Zon and Sparhawk do give a figure for 1922 of 9.87 m ha, but this includes 2,592,000 ha of forest in the northern territories, which is 'savanna forest'. Properly subtracted, this leaves 7,278,000 ha. Furthermore, even this corrected figure includes the areas which were inhabited and farmed.

Table 4.2 brings together a number of estimates of the area of closed forest made early in the twentieth century. It indicates that the area of the closed forest zone was then considered smaller than it is today (c. 6.6–7.2 m ha), and that c. 2 m ha of land within the closed forest zone was even then considered to be farmland. A straightforward subtraction suggests that the area of 'intact forest' around 1922 was perhaps as much as 5 m ha but not more; only about half of what Gornitz asserts.

That the extent of forest cover was exaggerated even in these early statistics was suggested by Meniaud (1933). He explicitly criticised earlier analysts of the forest zone for failing to distinguish between areas of the forest zone and areas within it which actually then carried forest. He suggested that in Ghana the area of 'Grande forêt' (closed forest) was only 4,500,000 ha and that even this figure 'must be reduced as in the high forest, there are patches of high grasses and low bush of savanna type (of which the natural cause is not always evident); considerable areas have been completely brushed or opened by axe and fire for yam, maize, manioc, and often oil palms and kola trees are almost the only trees left standing' (1933: 537).

In particular, Meniaud noted that the forest of the whole eastern region was very 'cut into' and estimated that 'one can only count on 2,500,000 ha as veritable intact primary Grande forêt, or exploited only by loggers'

Table 4.2 Early forest cover estimates for Ghana (ha)

Source	*Chipp 1922b*	*Gold Coast Handbook 1925*	*Gold Coast Handbook 1928 (2nd edn)*	*Gold Coast Handbook 1937 (3rd edn)*	*Report of the Forestry Department 1937/38*
Closed forest zone	7,257,200	7,224,000		6,601,644	
Merchantable forest	2,954,880		2,588,880	2,329,992	
Unprofitable or inaccessible	1,451,520		1,294,440	1,268,551	
Forest set aside for fuel	1,036,800				
Agriculture, fallow, plantation	1,814,000*	2,064,000 of which 390,000 cocoa		3,003,100	
Total forest	5,443,200	5,160,000	3,883,320	3,598,543	4,789,000

Note
*Estimated extent of farmed, fallowed and plantation land derived from population census data combined with estimates of land use per capita (Chipp 1922b).

(1933: 539). Given that other figures drawn up at the same period for similar forest types are at least 2 m ha higher, one should be sceptical of Meniaud's figure, exaggerated perhaps to show the endangered status of West African forests.[5] The figure must be an underestimate – but it does confirm the picture of large areas of the forest zone not being 'under forest'.

Early forest cover figures were themselves compared by early analysts to give the impression of precipitous deforestation. The third edition of the *Gold Coast Handbook* (1937) makes a calculation concerning the rate of deforestation:

> The process of shifting cultivation, the principal agricultural system in the country, is steadily eating into and diminishing the extent of the remaining forests. The average rate of forest destruction, estimated over a period of forty years [since 1897], and in the closed forest zone only, appears to have been approximately 75,000 ha/year.
>
> (*Gold Coast Handbook* 1937: 55)

This deforestation figure seems to have been derived by comparing the area of farmland estimated in 1936 for the *Gold Coast Handbook* 1937 with a similar estimate made by Chipp in 1920 (Chipp 1922b). This comparison indicated an increase in farmland – and hence supposed forest loss – of 1,189,100 ha over 16 years, a rate of 74,300 ha/year. This trend was then, apparently, backdated to 1897. Such a rate was seemingly supported by the figures for forest cover change over the same period (1920–36). The analysis thus implied deforestation of 3 m ha over the preceding 40 years, implying in turn that in 1897 there was no farming, merely forest, in the closed forest zone.

The area of forest in 1937 on which these pessimistic assertions were based was 3.5 m ha. Yet interestingly, this estimate was contradicted by the Report of the Forestry Department for the same year (1937/8), which calculated the area of forest in Ghana to be much higher, at 4.8 m ha. The importance of this slippage in the statistics can be gauged when Foggie (1953) estimated the area lost annually to farming between 1937 and 1953. Using the higher figure of 4.8 m ha for 1937 and an estimate of c. 3.5 m ha[6] for 1953, he calculated a loss of c. 70,000 ha/year. Had the lower figure for 1937 been used (the basis for the earlier assessment of deforestation) no forest would have been lost. As they stand, the two analyses imply that Ghana lost the same area of forest twice over. In a later article Foggie once again revised his deforestation rate upwards: 'If, in 1947, one estimated the wooded area at 4,377,000 ha . . . at the end of 1957, 2,747,000 ha . . . could still receive the denomination of forest' (1959: 11). His 'if' is a big if. The figure he gives for forest cover in 1947 is much higher than those accepted in the 1920s and 1930s. He goes on to state that whereas up to 1947 annual clearance stood at c. 65–75,000 ha/year, in the 1947–57 period the rate doubled to 150,000 ha/year.

In short, high rates and large extents of deforestation have been maintained by a combination of questionable extrapolation backwards of deforestation rates calculated between more recent dates, and statistical sleights of hand. But however spurious, the figures for deforestation calculated by early foresters were influential in their day, and have been accepted uncritically by many present analysts as properly reflecting early deforestation. The 1937 *Gold Coast Handbook*'s figures for deforestation rates were widely quoted at the time – Moor (1937) cited them to suggest alarm at the climatic implications for cocoa, for instance – and they were incorporated into the French literature via Gourou (1947) and Chevalier (1948a).

It could be thought that much of the farmed area in the forest zone in 1920 reflected immigration during 1900–20, perhaps linked to the cocoa boom of this period. In 1900 cocoa production was insignificant at around 536 tonnes, but by 1910 it had increased to 22,631 tonnes; by 1920, 124,773 tonnes; and by 1926–30 around 200,000 tonnes – after which it levelled off at c. 250,000 tonnes until c. 1950. Indeed, such cocoa expansion would seem to provide the rationale for the backdating of estimates of forest cover loss made in the 1930s to the turn of the century. Early foresters certainly did consider the forest 'crisis' to be linked to cocoa expansion; for instance:

> [M]any of these apparently forested areas are honeycombed with farms that are continually expanding, until now it may be said of the eastern half of this once wholly forested area that it is more farm than forest. This state of affairs appears to have been brought about since the introduction of cocoa, i.e., within the last 30–40 years. With the rate of expansion in the future being the same as it was in the past it is obvious that the coming generation will see the destruction of the remaining forests.
>
> (Moor 1924: 82)

But while it is certainly the case that large areas of forest were settled at this time by migrant cocoa farmers in the eastern region (see Field 1943; Hill 1956; Johnson 1964; Amanor 1994a), one still needs to ask to what extent cocoa expansion led to deforestation, rather than replacing existing land uses. How much cocoa production was carried out by populations already in place who had earlier been engaged in other forms of farming?

In Chipp's (1922b) estimate of forest cover, of the total land in the closed forest zone under agricultural production or fallow in shifting cultivation (1,812,000 ha), we should remember that only 11 per cent was attributed to cocoa; the rest was under food cropping and other more traditional tree crops. Gent, in 1925, estimated that of more than 2,000,000 ha under farming, less than 20 per cent was cocoa. Thus at this period of the cocoa boom, cocoa was by no means the major land user (Gent 1925).[7]

That much of the expansion in cocoa production was also by populations already in place was noted by Thompson when visiting Ghana in 1908, early in the cocoa boom. He noted that cocoa nurseries were present in all villages in Ankobra river basin, and also that:

> Nurseries of young cocoa plants were a constant feature of every Ashanti village that we passed through within the big forest region, and the same remark applies to the villages of Upper Wawsaw and Denkira. The way in which the cultivation of this plant has been taken up . . . throughout the most suitable localities in the colony and Protectorate is perfectly astonishing.
>
> (Thompson 1910: 76)

Thompson is noting a switch to cocoa farming by resident populations (albeit perhaps with some immigrant labour, sharecropping and independent planting), thus suggesting that much of the early expansion in the bulk of the closed forest zone was a transition in land use from food-centred shifting cultivation to home gardening with cocoa as the primary crop.[8] That the shift to cocoa early in the century may thus have been more an economic than an ecological transition is suggested indirectly by Wilks (1977), who argues that by the 1930s the food-farming sector of the rural economy was in a state of decline in many areas as a result of the increasing concern with the growing of cocoa as a cash crop.

To sum up at this point, many present analysts suggest that Ghana had around 8–10 m ha of forest around 1900, but data from early in the colonial period do not support this. The area of the closed forest zone was then considered to be only around 6.6–7.2 m ha, of which at least 1.5 m ha was already farmed.

Population and land use in the nineteenth century and before

To gauge the extent of forest loss during the twentieth century, it is also important to consider land use in the forest zone in the nineteenth century and before. If, as is suggested in most modern sources, significant deforestation began for the first time in the late nineteenth century, one must suppose that this region was hardly populated prior to this time, and that what is today defined as the forest zone was indeed a zone of forest in the nineteenth century.

Several foresters have argued that the forest region was hardly inhabited prior to the twentieth century. Hall (1987, citing Gaisie and de Graft Johnson 1976), for example, suggests that the population of the closed forest zone increased from about 250,000 in 1850 to 750,000 in 1900 to 5 million today, implying a more or less exponential rise. But this demographic analysis is now long superseded. Indeed, recent work by historians suggests that the nineteenth century was a time when populations in the forest region generally fell, rather than increased. The weight of the historical and archaeological record now suggests tentatively that processes of depopulation in Ghana's forest zone may date back several centuries earlier, and be linked to Ghana's experience of the slave trade.

One can begin with the work of Wilks (1975, 1993) on Ashanti (see map 4.3). Merely considering 'Metropolitan Ashanti', a square zone of c. 1,657,000 ha centred on Kumasi, he estimates on the basis of the contemporary estimations of Bowdich and Freeman that the population in the early nineteenth century was 500,000–725,000. He suggests that during the nineteenth century Metropolitan Ashanti's population declined to 250,000–375,000 (1975: 87–93). While the early Ashanti population has been the subject of debate (see Wilks 1975, 1977, 1978; Johnson 1978, 1981; Klein 1994, 1996), the state of the forest at the turn of the twentieth century, as we indicate below, would give Wilks' higher figures some support. He goes on to estimate, on the basis of the population combined with estimates of land use per capita under shifting cultivation, that with the lower population figure for 1800 (500,000), around 600,000 ha were farm or fallow. This estimate may well be conservative. First, were the higher population estimate (725,000) to be accepted, it could be raised to

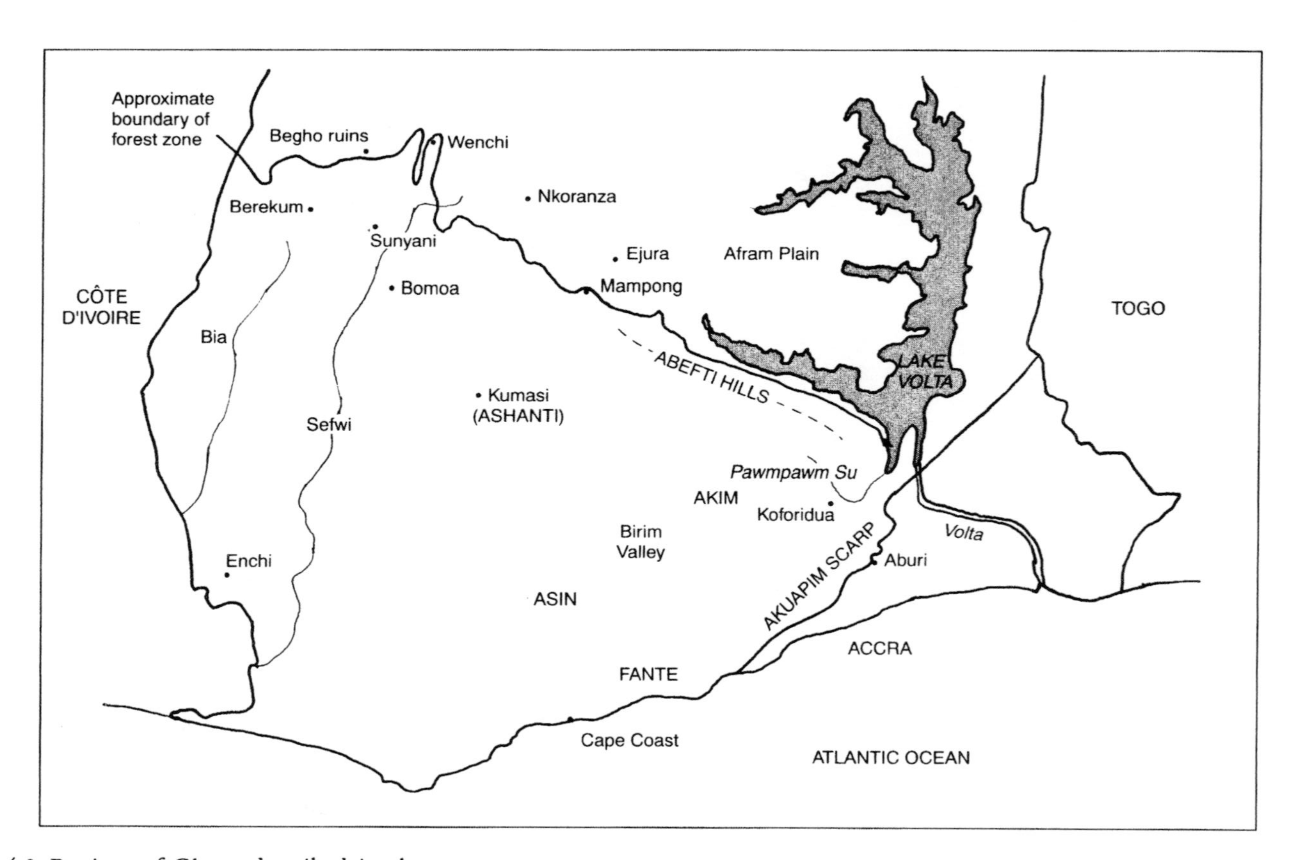

Map 4.3 Regions of Ghana described in the text.

870,000 ha. Second, village sites frequently move and fresh areas are brought into cultivation. Thus a large proportion of land not then actually within fallow cycles might well have been recently farmed. As Thompson noted: 'The result is that large tracts of forest-covered country, quite out of proportion to the inhabitants they have to support, become involved in the general purpose of destruction, and this is brought about by quite a small population' (Thompson 1910: 33). Many early European visitors romanticised the 'forest' vegetation of Ashanti, but not all. When Huppenbauer visited Kumasi in 1881, for example, he noted that 'the actual land of Asante is not forest, as Akyem for instance, but mostly cultivated' (Wilks 1978–9: 52).

The decline in Ashanti populations seems to have been a dominant trend throughout the nineteenth century but it certainly accelerated from 1863 to 1911, years of military campaigns, civil war and displacement. As the British governor commented in 1891:

> [A] part of Koranza, and also a part of Mampon, together with Dadiassi, Kokofu, and Inquanta, all powerful tribes, have crossed from Ashanti and sought refuge in the British Protectorate, and the countries they have left are being rapidly overrun by bush and forest, farming and trade operations having ceased in them. . . . Adansi also, is without population. The country is fast becoming forest.
>
> (cited in Wilks 1975: 91)

It therefore seems that the population of Ashanti alone, let alone other areas of the forest zone, was around 350,000 in 1900 and that earlier in the nineteenth century this figure would have been larger, not smaller. Parts of the 'forest zone' around 1900 were, therefore, covered with forest regenerating on lands depopulated during the nineteenth century or before, whether through death or migration. Land settlement early in the twentieth century might partly have represented the return of populations to land which they had vacated some years before, as well as new immigration.

Wilks focuses on only a small part of the forest zone. It is worth examining briefly the population and farming history of other areas. Here we take three examples: the area around Enchi in Western Ghana and, further east, Asin and the Birim valley.

Enchi

Western Ghana has a richly documented history, including one of depopulation during the eighteenth century. Prior to the expansion of the Ashanti state in the eighteenth century, much of Ghana's forest region had already fallen under the control of the earlier Denkyira state which Ashanti defeated in 1701. At the time of its fall, Denkyira incorporated six kingdoms; one of these, Eborosa (otherwise known as Aowin) was in south-west Ghana with its capital at Enchi, and main towns at Yakasse, Kwahu and Jema.

Prior even to the rise of the Denkyira state, Aowin (Eborosa) had been the most powerful state in Ghana's south-western region. It was not until the last two decades of the seventeenth century that Denkyira gained control of Aowin and its trade routes to the emergent southern ports (Daaku 1971: 33 in Perrot 1982: 48). Denkyira defeated Aowin in a bloody war (Bosman 1705: 79). Subsequently the defeated Aowin became repeopled by taking refugees. With the fall of Denkyira to Ashanti, Aowin expanded its control to incorporate Sefwi to its north.

The Aowin were then attacked and defeated in 1715 by Ashanti. They nevertheless remained powerful enough to muster an army of 8,000–9,000 people to rebel against Ashanti, capitalising on the political vacuum caused by the death of the Ashanti leader, Osei Tutu. Led by Ebiri Moro, the Aowin were even able to sack Kumasi, but were then routed during their flight. This defeat gave Ashanti the land between the Tano and Bia rivers, and the population of this region was crushed in revenge. As Fuller described, this tract of land became the game reserve of the Ashanti Kings, the New Forest of the conquerors (Fuller 1921). Following defeat and Ashanti retribution, many Aowin left the Enchi area for a region in what is today within the proposed Bia Tawya reserve, centred on the ancient fortified town of Konvi-Ande. Subsequently they moved once again, to the west where their descendants are called Anyi (Perrot 1982). Enchi, once a thriving state, became depopulated.

One might be tempted to suggest that despite this history of warfare, the level of eighteenth-century Aowin populations had little lasting impact on the forests of this region. Yet the error of such deductions is clear from the description of Thompson, who, during his 1908 survey of forests, visited the area west of the Tano river, to the north and north-east of Enchi:

> I was rather disappointed in these forests as we were led to understand by the guides that they were extensive and practically virgin in character. This we found to be very far from the case, and the whole tract of country showed unmistakeable signs of villages, having been once pretty well inhabited. Large tracts of forest were found to be of secondary origin, and signs of villages having once existed here were also not wanting. In fact, on our return to the village (Tomento, east bank of Tano) the chief admitted that a very long time ago the country had been inhabited by a people who had since moved westwards.
>
> (Thompson 1910: 46)

It is thus a sign of enduring depopulation, including emigration from the area, that we read in the Report on the Department of Soil and Land Use Survey (1956: 4) that 'the Sefwi-Bekwai and Asankrangwa areas are relatively densely populated and closely farmed, but population is sparse in the Enchi area and much untouched forest remains'. There may indeed have been much forest, but the area's demographic history denies its untouched nature.

Asin

When Dupuis passed through the Asen (Asin) region on his way to Kumasi in 1819, a region to the west of the Birim valley, he noted the state of depopulation there. While interpretation of such early accounts obviously requires caution, it is worth repeating his analysis in full, as it clearly sets the terms for any debate.

> It will not be deemed extraordinary . . . that all Fantee, excepting the few towns on the coast, and the far greater portion of Assin should be found in the state I have described; the population extinct, the plantations more or less destroyed, and the forest relapsing to its original growth. In the wars . . . a few of the subjugated are reserved to grace the victor's triumph; but slavery, or death, is most frequently their final doom. Little kingdoms have been thus annihilated, as in the case of Denkera, Akim, Warsaw, and many others, whose names have almost become obsolete. The importance of these dreary regions is, therefore, trifling, when compared with a surface of such extent; and particularly so, when it is considered that where a population amounting, it is supposed, to something between three and four millions occupied the districts of Fantee and Assin, the inhabitants now scarcely amount to as many hundred thousands, and after the most liberal calculation, including even the Fantee towns on the sea side.
>
> (Dupuis 1824: 45)

While Dupuis noted that by 1819 the forests were recolonising this region, he nevertheless estimated the populations of Fantee and Asin at the time to be c. 300,000–400,000. These people lived in what is today described as within the closed forest zone, whether characterised as southern marginal, dry semi-deciduous, moist semi-deciduous or moist evergreen. Added to Wilks' estimate of the 1800 population of Metropolitan Ashanti as 500,000–750,000, this region alone brings the population of the forest zone up to, perhaps, over 1 million.

Birim Valley

As Kiyaga-Mulindwa (1982) notes, authors such as Hill (1956), and Hunter (1972) argued on the basis of oral tradition that the Birim valley was uninhabited until its settlement by the Atweafo in the seventeenth century. Kiyaga-Mulindwa, however, disputes this. Archaeological finds suggest that an earlier population had been destroyed in what he terms 'slavery-induced ethnocide'. He argues that this ex-population was culturally continuous with those inhabiting the forest region from the first millennium AD, whose Iron Age economy was based on forest cultivation and hunting. He argues further that this population was perhaps continuous with the Neolithic and Early Iron Age populations of the area in the period between 1000 BC and AD 500, a deduction supported by Posnansky (1982) and Klein (1996). The ex-population inhabited the whole of the western Densu Valley, all of the Birim Valley and the western part of the Pra Basin, probably as far south as the Kwisa Hills and the River Ofin, north to the slopes of the Kwahu Mountains and on the Accra plain to the south-east. Mulindwa argues that these populations established defensive earthworks in the period immediately preceding their disappearance in the sixteenth and seventeenth centuries. The earthworks would have been sufficient defence against the persistent but limited pressure from outside during the early years of the slave trade, but as this developed and used new armaments, the fortifications became inadequate. Thus the majority of the early Birim Valley populations were targeted for capture and export as slaves over a period of a generation or two between the seventeenth and early eighteenth centuries. Material evidence indicates that no significant remnant of the earlier occupants of the Birim Valley was left when the newcomers, the Atweafo, resettled the valley, finding it open to their entry. Subsequent warfare and slave trading prevented populations in this region from recovering their former levels until the twentieth century.

Both Isert (1788) and Riis (1842, in Johnson 1964) suggest that the kingdom of Akim suffered the same fate of depopulation. Isert, for instance, noted that 'the Akims must have been a highly populated country in earlier times, before they were subjugated to the Assianthees' (Isert 1788/1992: 173).

From such evidence, it would seem to us that populations in Ghana's forest region prior to the twentieth century were far higher than credited in most forestry literature. In many cases these higher populations were perhaps more relocated than destroyed, so each instance of depopulation in one region implied another of repopulation elsewhere. But that 'elsewhere' was often the Indies and Americas. The impact of the Atlantic slave trade on West Africa in general has been the subject of strong debate, at times heated (see Curtin 1969; Lovejoy 1982, 1983, 1989; Henige 1986; Eltis 1987; Manning 1992). Work which examines its demographic impact must use figures and opinions concerning the quantity, extent, timing, ethnic, sex and age composition of slaves, and the related wars, disease, refugee fluxes and famines. With so many variables, the debate inevitably remains inconclusive. In sum, however, simulation of its demographic impact suggests an absolute African population decline, not just a decline relative to what there would have been otherwise (see Lovejoy 1989; Manning 1992). As Lovejoy sums up: 'the size of the trade, including enslavement, related deaths, social dislocation and exports, was sufficient to have had a disastrous demographic impact' (1989: 393). While Ghana may have prospered in the fifteenth and sixteenth centuries, the coincidence of the Atlantic and interior slave trade with access to new weapons and interior warfare for diverse reasons may have meant an immense negative impact on forest populations.

The pre-sixteenth-century population history of Ghana's forest zone remains the subject of some controversy. Wilks (1977) argues that there was an era of massive land clearances dating from the fifteenth to the seventeenth centuries when a switch from hunter-gathering to food cropping took place. He bases this argument on a reading of Ashanti oral history in keeping with the position of early historians who considered the forest region to have been almost, if not quite, uninhabited prior to the coming of Europeans (Ward 1958; Bourret 1960; Apter 1963). Several Ghanaian historians – such as Adu-Boahen (1977) – have strongly denied the latter point. Yet other evidence questions Wilks' analysis. First, it appears that the gold and kola trade stimulus to which Wilks partially attributes economic settlement of the forest zone post-fifteenth century occurred earlier than Wilks supposed, possibly c. AD 1000 (Posnansky 1982). Second, there is evidence that while oral accounts may describe a movement into uninhabited forests, these forests may have been depopulated a short time before, and depopulated by long-established farmers.[9]

Recently Klein (1994, 1996) has further questioned Wilks' proposed demographic history of the forest, arguing on the basis of the distribution of sickle cell trait and archaeological evidence that 'agricultural ancestors of present-day Akan populations have inhabited the forests of southern Ghana for approximately two millennia' (1996: 264). He argues that 'the sama data which have been used to prove a "population explosion" during the sixteenth and seventeenth centuries actually indicate . . . [a] clustering response on the part of survivors, struggling to escape the social and political dislocation which followed in the wake of the Atlantic slave trade' (1996: 249).

Those who consider the forest region to have been hardly inhabited prior to the cocoa boom thus overlook the evidence that the forest region held considerable farming populations in the mid to late nineteenth century, and that populations around 1800 and before may well have been larger, not smaller. Of a forest zone of around 7 m ha, possibly 1.5–2 m ha were farmed around 1880, and more earlier in the century. Furthermore, it is quite possible that the twentieth century is seeing the second clearance of Ghana's forest, the first having taken place from c. AD 1000 to 1600. Arguably, there may be more forest in Ghana today than in the early seventeenth century.

Today's forest reserves at the turn of the century

The extent and impact of early populations of the forest zone at the turn of the century and the legacy of depopulation can be gauged from Thompson's study of Ghana's forests in 1908. Conservationist authors who like to consider Ghana's deforestation as recent claim that Thompson visited Ghana in response to a perceived need for conservation 'due to *recent* farming pressure on the forest'; and that he described 'alarming deforestation'. Yet as the first two paragraphs of his report make clear, he was invited to report on the forest resources of the country and the measures that should be adopted for the preservation of the forests *against excessive timber exploitation*, and to regulate 'the haphazard methods of exploiting the mahogany forests then in vogue' (1910: 5, our emphasis). So while his findings did spotlight the 'problem' of shifting cultivation, this was not construed in terms of 'recency' and it was certainly not the *raison d'être* for his visit. Equally, while Thompson proposed reserves, he was on his journey also evaluating the potential of areas which others had earlier suggested as forest reserves. As we shall see, he often rejected such sites, finding them to be either recent secondary forests with little timber value or extensively inhabited and farmed.

Thompson's report certainly does not give the impression of entering a recent forest 'frontier'. Indeed, Thompson had to go looking for forests. As he wrote of his choice of route (1910: 6): 'It was more important to discover what forests were left intact, and to explore wooded areas about which but little was known.' This comment alone suggests the extent to which Ghana's forest zone was not high forest in 1908.

In many southern regions of the forest zone, Thompson observed old forest only on ridges and hills, or in patches 'here and there'. The forests south-east of Kumasi, up to and around Lake Bosumtwi, and east and south-east of this were largely farmed. Such forest as there was, generally was either of small extent, old secondary forest on territorially disputed land, or on hill ranges. The only areas where there were extensive tracts of forest were south-west, west and north-west of Kumasi. That he did detail several extensive tracts of forest only serves to underscore the relative scarcity of such tracts. It is true that large parts of these have since been lost. But it is equally true that a large proportion had already been lost long before, if they ever existed.

That all the forest which could be reserved was reserved is also suggested by a number of other factors. These include the irregular shape of existing reserves, presumably determined by the need to maximise the reserved land; the extent to which populations were forcibly displaced from reserve areas, or existing inhabitants guaranteed rights to continue their farming in delimited enclaves within reserves; and the large number of small forest reserves: of about 225 reserves 25 per cent are less than 1,000 ha, and 56 per cent less than 5,000 ha.

Thompson's report also underscores how many of today's high forest reserves were, when originally reserved, only recent secondary forests:

> When selecting forests for reservation, it will be found that comparatively few tracts are covered with so-called primeval or virgin forest; the majority of forests on the Gold Coast and in Ashanti consist of secondary irregular growth that has sprung up on areas previously cleared for farms by the natives. In places, such forests have, since they re-occupied the abandoned farms, been left untouched for such a long time that a sufficient interval has elapsed for the trees composing them to have grown into large trees of very nearly as good growth as the original ones that were felled. Such forests should be selected in preference to those of younger age, commonly met with closer to the larger native towns and villages.
>
> (Thompson 1910: 147)

Thus it is clear that the condition of Ghana's forests early in the twentieth century prior to their reservation was anything but pristine. On more than 20 per cent of the land which eventually became reserved, the reserve working plans attest to the presence of significant past populations. Yet it is against the image of pristine forest – not the state of forest when reserved – that today's forest condition is assessed (see Hawthorne and Musah 1995).

Hawthorne and Musah state that many northern forest reserves had no forest in them when reserved (1995: 13). Nevertheless, in a detailed analysis to inform both the Ghanaian forest department and then ODA projects in the region, they define and map modern 'forest reserve condition' only in relation to the forest which could be supported there, not in relation to what forest was there (or not) at the time of reservation. This conventional practice in forestry science gives a strong impression of the erosion of Ghana's forest resources, especially in the northern fire zone reserves, all described as 'highly degraded' or with no forest left. But many of these reserves were not forest in the first place. It would seem more appropriate to indicate the state of the reserves relative to their state on reservation, rather than in relation to a supposed 'original climax vegetation'. Table 4.3 – which risks presenting rather suspect figures if only to provide order-of-magnitude estimates – summarises the different figures produced by these two types of comparison.[10] Of the 0.9 m ha of forest reserves today classified

Table 4.3 Estimated change in status of Ghanaian forest reserves classified today as in 'unsatisfactory condition'

Classification of degradation of forest reserve	*Area of the reserves so classified*	*Estimated area in this state prior to reservation*	*Actual 'degradation' of the forest reserves*
No forest	240,100	240,100	0
Very poor	262,600	211,900	50,700
Mostly degraded	369,700	98,600	271,100
Total	872,400	550,600	321,800

as in unsatisfactory condition, more than 60 per cent were probably in a similar state prior to reservation.

Issues at the forest margin

Until now, we have briefly examined population and land use history in 'the forest zone' without clearly identifying what constitutes that zone and its extent in the Ghanaian context – although we have mentioned that today's analysts generally consider the zone to be larger than did their 1920s predecessors. While vegetation zonation is clearly a useful device for examining the spatial variation in existing vegetation, many analysts of Ghana – as of other countries – have been tempted also to deduce that zones deemed (on bio-climatic grounds) to have a particular vegetation potential actually carried their potential vegetation until they succumbed to 'anthropogenic degradation'. Yet as noted in Chapter 1, the notion of 'original forest' can be questioned from several angles. Episodes of climatic change, natural fires and elephant impact, among other factors, must have interacted with populations over thousands of years, forcing the concept of 'originality' to be questioned.

As we shall now examine in detail, large parts of the area which Hall and Swaine (1981) suggest are part of the closed forest zone (and hence which conservationists construe as having been deforested) were certainly not classified as such by early foresters and travellers. These issues of zonation are most relevant on the forest margins, where many authors have suggested that large areas of present savanna could support forest, and did 'originally' until forest was lost to 'derived' savanna. On this basis Sayer *et al.* (1992) suggest that the closed forest zone 'originally' covered some 14.5 m ha. This view is in keeping with those of most early foresters, who also assumed that the Guinea savanna zone would be largely forest covered were it not for the activities of farmers, and that these savannas were thus derived (Chipp 1922b, 1927; Gent 1925; Moor 1936). As Hall (1987) notes, belief in the derived savanna hypothesis during the 1920s and 1930s, and of the climatic impact of deforestation, was almost religious. But while the savannisation theory has dominated much reasoning concerning vegetation change in the transition zone, historical evidence which can be assembled for the last 100 to 200 years suggests very different conclusions concerning forest–savanna dynamics.

Here we examine the centre-west transition zone in the Sunyani–Wenchi area and the eastern/south-eastern transition by way of examples, and then look briefly at the area east of the Volta.

Centre-west transition

That vegetation in the transition zone of the centre-west region reflects the ongoing degradation of forest to savanna has been accepted by several authors (Taylor 1952; Asare 1962). Taylor, for example, considered that the forest once extended up to 8.30° N. Asare notes that: 'It is likely that the area referred to as transition zone . . . was once occupied only by high forest. Owing, however, to biotic (mainly burning and cultivation) and edaphic factors, some parts have been replaced by derived savanna' (1962: 68). He suggests that what forest remains does so because it is less cultivated.

Intriguingly, Asare (1962) did note several areas in which savanna had become forest. He suggested that 'when derived savanna is left uncultivated and unburnt, it grows back to forest', and that it 'passes through several stages of development, the most distinctive formation being transition woodland'. Thus his analysis of new forests was contextualised more broadly as a stage of 'recuperation after degradation'. More recently, several authors such as Swaine *et al.* (1976) have questioned Asare's analysis of this part of the forest–savanna transition, suggesting that the vegetation appears to be more stable and that, even in settled areas, the pattern of forest and savanna is determined largely by non-anthropogenic factors such as soil, slope and hydrology. Swaine *et al.* question Asare's (1962) conclusion that savannas are the result of farming, pointing out that forest patches tend to be more heavily farmed than savannas. Swaine cites Spichiger and Blanc-Pamard's (1973) conclusions in neighbouring Côte d'Ivoire that thicket, not savanna, becomes established on abandoned savanna farms. In other words farming can enhance the presence of secondary thicket in savanna, rather than degrade existing thicket to savanna. Markham and Babbedge (1979) also support this analysis of the centre-west transition. They note that of the hilltop forests, parts contain numerous heavily overgrown mature specimens of savanna trees, suggesting that the area may previously have supported savanna (1979: 228).

While these ecological studies of the fire zone rightly question the assumptions driving the derived savanna hypothesis, they nevertheless overlook important issues concerning past habitation and land management. First, both Swaine *et al.* (1976) and Markham and Babbedge carried out their ecological analysis on a site which was adjacent to the most significant ruined town in Ghana, Begho, which has eleventh-century origins and was, at its height in the seventeenth century, a major centre for weaving, iron smelting, copper working and pottery (Anquandah 1992). It seems odd that the vegetation could be examined without reference to the site's history. More recent history is also relevant. On visiting the site, one finds that the forest sections of the transects studied by Swaine *et al.* fall within part of the anthropogenic peri-village forest island of an abandoned village called Deke. Inhabitants left this village around 1900 to settle in the savanna at the present site of Degedege. Subsequently, around the new site an extensive peri-village forest island has also formed, in which *Antiaris toxicaria*, *Triplochiton scleroxylon* and *Milicia excelsa* trees were abundant until their recent felling for timber.[11] It appears that the forest patches in the region might be intimately linked to past habitation and land management, if not the result of it. Swaine *et al.* and Markham and Babbedge all describe as natural a forest vegetation which exists over ruined villages, and is generally characteristic of these being especially rich in *Anogeissus leiocarpus* (see Keay 1947; Sobey 1978).

The need to examine vegetation from a more historical perspective extends in this region to the analysis of sacred groves. There is a tendency to consider these as oases of 'indigenous nature conservation' in an otherwise unspared landscape. Indeed, conservation programmes often treat these groves explicitly as relics of the region's former natural forest vegetation. Yet frequently these groves cover the sites of old settlements, and as such their vegetation can hardly be considered as 'natural'. In the transition zone the forests are very likely anthropogenic formations. The grove of Pepease, near Berekum, is a case in point. Considered a natural relic by conservation programmes, this forest actually covers the old capital of Berekum and remains the burial place of its chiefs (Ahin 1973: 3).

Recent analyses, whether of derived savanna or stability, have also overlooked or ignored other evidence that the region has been ceding from a predominantly savanna landscape to a predominantly forested one in historical times. Indeed, and somewhat paradoxically, such analysis of forest advance seems to have been the received wisdom even in forestry circles prior to independence. The Chief Conservator of Forests, Foggie (1953), noted the advance of the forests in this region: 'In the north-west, the savanna at one time extended much further south. The forest reserves north and west of Sunyani are rapidly changing from savanna woodland back to closed forest' (1953: 132). While Foggie attributed this forest advance to depopulation following warfare in the early 1800s, an earlier colonial forester, Vigne, who had noticed cases of closed forest encroaching on 'savanna forest', attributed it to climatic change, considering that the Gold Coast was at the time experiencing a wet cycle, as indicated by a rise in the water level of Lake Bosumtwi:

> In tension zones, relatively small climatic changes may have important influences, and it is difficult to account for the large extent of 'savanna forest' in the area I studied by assuming a larger population in the past; I consider it is due partly to drier climatic conditions in the past, specially as measured in rainfall and humidity over the fairly short dry season.
>
> (Vigne 1937: 93–4)

That the forest was advancing in this region was hypothesised by one of its earliest 'explorers', Freeman, who passed through it when walking from Kuamasi to Bonduku in 1898.

> The forest near Fatenta (west of Sunyani) was different from that further south. Although dense, lofty and umbrageous, there was an absence of those immense patriarchal trees that formed the bulk of the forest in the central region, and most of the trees of the silk-cotton order were quite young. It appeared to me that the forest was extending in a northerly direction, for not only were the trees, as I have said, mostly young, but for several days at increasingly frequent intervals, we had crossed patches of woodland covered with the small contorted trees characteristic of the country north of the forest, thickly interspersed with young members of the silk-cotton tribe.
>
> (Freeman 1898: 164–5)

The hypothesis of forest advance into savanna regions was also suggested by earlier accounts. Dupuis (1824: 28) noted that, travelling from Kumasi to Gayman (Bonduku), one would reach 'open plains on the Tano river'. Bowdich (1819: 169) wrote that Buntookoo, the capital of the kingdom of Gaman, was '11 journies N.N.W. of Coomassie. . . . The river Ofim is crossed the second day, the Tando the fifth [i.e. at c. 114 km, just south of Odumase-Sunyani], thence the country becomes open'. The earlier presence of savannas some 40 km south of Sunyani is suggested in the oral histories of the Dormaa people, whose centre in 1700, it is said, lay in grasslands at Abanpredease [now Bomaa] near the Tando river (Owusu 1976: 6–7). References to areas today considered as within the closed forest zone as savannas are found in even older documents. For example, the celebrated schematic map of 1629 (reproduced in Anquandah 1982) described the kingdom of Bono (with its capital at Bono Manso near Techiman) as 'of simple people and no forest'. Dapper wrote of the Kuiforo kingdom immediately south of the Bono kingdom – into the forest zone – that 'one finds no bushes there at all' (Dapper 1670/1967).[12]

In sum, one would be hard pushed to suggest that the history of vegetation in this part of the transition zone during the twentieth century has been one merely of deforestation and savannisation. Of the many forest patches known to have declined in quality or area, one needs to ask about their origins, rather than assuming them to be 'natural' formations. While the zone north of Sunyani is today considered to belong to the fire zone subtype of Ghana's dry, semi-deciduous forests[13] (Hall and Swaine 1976), there is sufficient evidence to suggest that this has not always been the case. If Vigne's analysis is even partially correct (and comparative analysis westwards in Côte d'Ivoire and Guinea would suggest that it is), then it would be incorrect to classify this region as part of any 'original' closed forest zone.

Amanor (1993) recently reviewed the arguments for savannisation, stability or forest advance in the Wenchi area, and concluded that the forest vegetation was under threat. In the territories of many transition zone villages, there does appear to have been a recent loss of forest cover. Cocoa under a forest canopy has in many areas been converted to land for maize and other crops, apparently for both ecological and economic reasons. A huge drought-related fire passed through the region in 1983–4 and farmers have found it less profitable to re-establish cocoa than to turn the land to maize. But this forest loss cannot be considered merely as part of a long-term trend. It may be more appropriate to consider it in relation to a particular conjuncture of ecological and economic variables, this time leading to loss, with other conjunctures at other times perhaps having led to gain.

Eastern and south-eastern transition

Probably the best case for savannisation at Ghana's forest edge has been made for the eastern transition zone. The region in question is basically the valley of the Pawmpawm Su river, where the hills to the west (scarp range) and those to the east (along the Volta) almost join at the southerly end near Koforidua, forming a funnel which concentrates the prevailing harmattan wind. The forester Moor, an avid proponent of the savannisation–desertification model, visited the region c. 1929. He noted that:

> There is little forest left, but from the evidence of the odd trees and patches remaining, it is obvious that, up to very recent times, this part of the country was covered with a deciduous type of forest of good average height and density, with only a narrow transition belt between itself and the grass savannah. Now a belt of oil palms created some 50 years ago, neglected, has replaced the forest east of Pawmpawm Su; the greater part of the transition belt has been absorbed by the savannah, which has also thrust a wedge up the valley of the river; and the rest is a mixture of good and bad cocoa farms, areas of oil palms and sporadic patches of savannah and secondary growth, with here and there the remains

> of the original forest. Most of the grass patches lie more or less along the axis of the funnel, culminating in a final patch at the top of the Scarp range at the very source of the Pawmpawm Su. From the ages of the dry zone species now established in these patches, it is clear that they are of recent formation. This savannah intrusion is quite definitely the direct result of excessive deforestation.
>
> (Moor 1936: 2–3)

Cocoa was introduced into the region in 1908 and, Moor states, the forest was removed by 1920. By 1925–6, he argues, the cocoa plants had become deciduous due to drought and, this being coupled with swollen shoot disease and a major dry-season fire (1927–8),[14] cocoa production declined. In short, he argues, the country had twenty years earlier been under a forest which had disappeared (Moor 1936 [1929]: 2–3).

Yet prior to Moor's dating of the onset of savannisation, there was already savanna grass on these hills. This was noted explicitly by Thompson during his 1908 visit: 'I am sorry to say that a great deal of damage has already been done to these [hill] forests . . . as, for example . . . along the slopes of the Abetifi hills facing the Afram plain, the slopes of the Aburi hill system facing the Prampram and Accra plains and, so I am informed, of the slopes of the hills adjacent to the Krobo plains. Grass penetrates these hills' (Thompson 1910: 97). Moor's savannisation analysis was criticised by soil scientists who, surveying the region in the 1950s, considered the savanna penetration to be due to soil conditions.

> Settlement has gone on over the past 40 to 50 years and the expansion westward of the Krobo farmers is still proceeding rapidly. As a result of the Krobo farming system, little forest now remains in the region except on inaccessible slopes on the Kwahu scarp and the Akwamu hills near Jaketi. Savannah vegetation occurs in the northern part of the region as far south as Anyaboni. This does not, in fact, appear to represent an encroachment of savannah into the forest zone since the former vegetation is practically confined to the groundwater laterites over mudstones and to shallow soils over quartzite in the Akwamu hills.
>
> (Department of Soil and Land Use Survey 1956: 7)

More recently Amanor (1994a) has accepted Moor's (1936) savannisation analysis, considering that by the 1930s cocoa was devastated by swollen shoot disease, the land savannised and the landscape transformed by the food-farming system of immigrant Krobo farmers. Amanor goes on to show superbly how present Krobo farmers enrich the productivity of their land

and the woodiness of their fallows, but suggests that this is a phase of innovative regeneration following the earlier savannisation.

Thus Amanor argues that a series of distinct micro-environmental belts run from pure grasslands in the east to forest fallows in the west. He examines present vegetation patterns using a 20 km transect from an area in more or less open savanna (at Akruru Saisi) to an area within forest (Odometa). The savanna, he argues, was derived from forest, and 'vestiges of the forest seed bank still remain, and rare forest trees can often be found incongruously in savanna scrub' (1994a: 78). Once again, 'relict forest trees' are taken as evidence of deforestation. This land was settled between 1890 and 1930; following Moor's analysis, he suggests that the frontier was wealthy in cocoa during the early years, but went into decline during the 1920s.

Yet maps from the turn of the century indicate that the vegetation prior to dense settlement at the central and eastern part of Amanor's transect was 'grass plains and patches of forest', not forest as Amanor contends (Map 4.4). It may be that vegetation differences along the transect which Amanor took to represent different degrees of degradation are actually more stable vegetation forms; areas currently undergoing anthropogenic regeneration may not have been degenerated in the first place.

Two further questions need to be asked. First, were the forest patches which existed in this savanna linked to the ruined village sites of earlier inhabitants? There are, as Johnson suggests, traditions of a war between the Jaketi people (once inhabiting this Asessewa region) and the present Begoro people (to the west) in the mid-eighteenth century which led to the expulsion of the Jaketis (Johnson 1964: 11). Farmers in the region distinguish the more fertile and sherd-ridden soils of old settlement sites from normal farmland; is there a tendency for these sites to carry thicker forest?

Second, what was the area's vegetation like prior to the boom in oil palm prices and production in the 1860s? It is usually accepted that Krobo oil palm plantations were established on land cleared from forest. But as Aubréville and Chevalier noted in Benin, oil palms can be established on farmed savannas in the humid savanna zone, as the period of cultivation provides sufficient early fire protection. That Krobo farmers may have been doing this is suggested by Schrenk, who in 1865 told a Select Committee that: 'There is a large plain near the Volta which once had no trees on it about 100 years ago, and the whole of that plain is now a palm forest . . . planted by the natives' (in Johnson 1964: 21). Equally Zimmermann in 1867 reported that the 'Krobo buy the uncultivated but still cultivable land from the people of Akuapem, who do not cultivate it. . . . The Krobos build plantations, especially palm forests' (in Wilson 1991: 202). Later that year, Reverend Aldinger wrote: 'Very diligently they [the Krobo] extend the cultivation of oil palm year after year, thus building up oil palm forests' (Wilson 1991: 78). The phrase 'building up' would seem to suggest raising

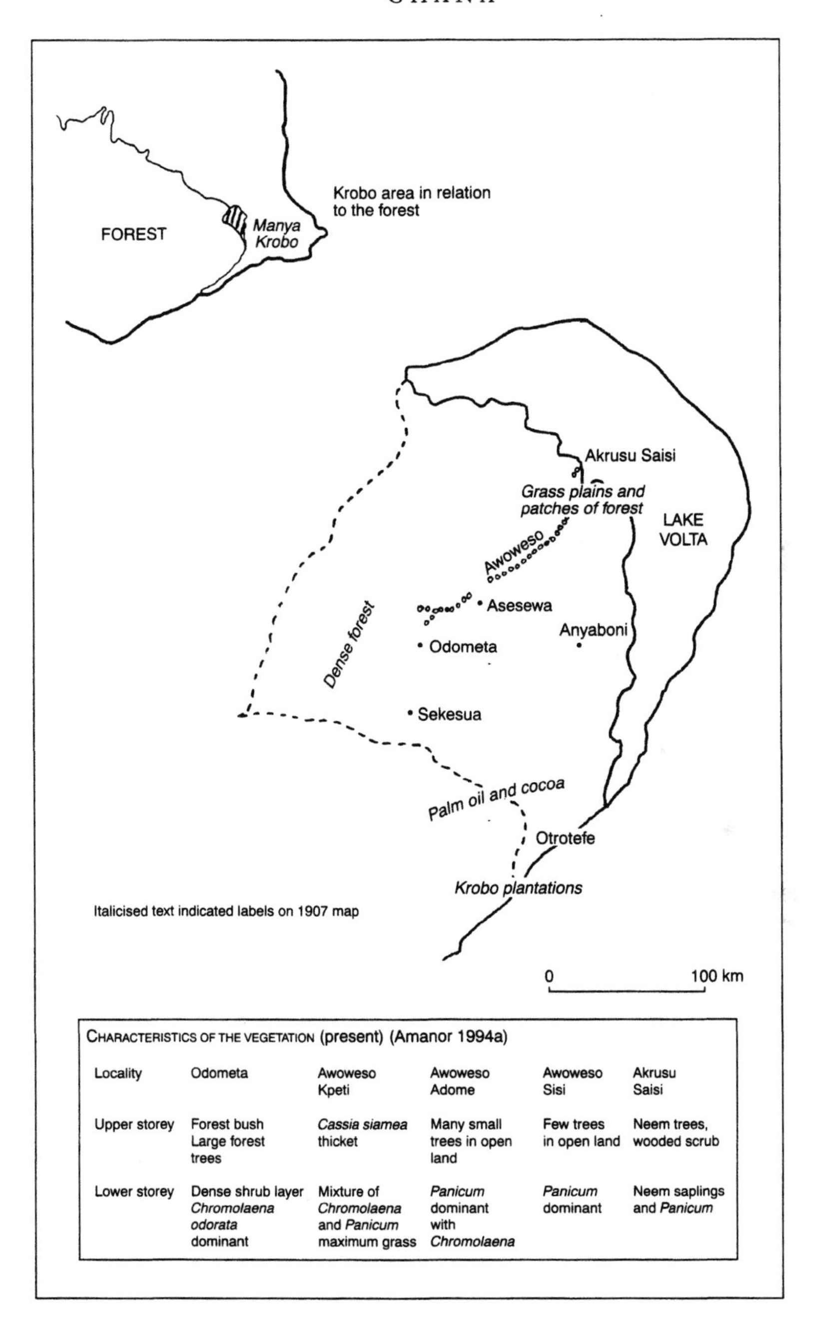

CHARACTERISTICS OF THE VEGETATION (present) (Amanor 1994a)

Locality	Odometa	Awoweso Kpeti	Awoweso Adome	Awoweso Sisi	Akrusu Saisi
Upper storey	Forest bush Large forest trees	*Cassia siamea* thicket	Many small trees in open land	Few trees in open land	Neem trees, wooded scrub
Lower storey	Dense shrub layer *Chromolaena odorata* dominant	Mixture of *Chromolaena* and *Panicum* maximum grass	*Panicum* dominant with *Chromolaena*	*Panicum* dominant	Neem saplings and *Panicum*

Map 4.4 South-east transition zone of Ghana: present vegetation transect (Amanor 1994a) compared with vegetation as described in a map of 1907 Gold Coast, 1:125,000 Komfrodua, 72. L.IV. Director of Surveys, Oct. 1907.

palm forests from lower vegetation; if this was uncultivated, it is unlikely to have been low fallow bush, and more likely to have been savanna. By 1891, a mission magazine described the isolated compound farms associated with Krobo agricultural expansion as lying 'in the palm forest, under mango and lemon bushes'. Croft, who walked from the Volta to Odomase in 1872, passed through 'groves of palm trees nearly the whole distance' (Croft 1874: 189). These are not forests with palms, but forests of palms. Certainly the model of oil palm expansion into a virgin forest frontier needs to be demonstrated for each locality, not assumed. And where cocoa was established 'in forest', could this sometimes have been anthropogenic oil palm forest?

Several recent researchers apart from Amanor have examined this zone and the region just to the south in terms of twentieth-century deforestation. Taking three case study sites,[15] Gyasi *et al.* (1995) argue that 'until about 1900, the southern sector of the zone was covered by unused or sparsely used closed deciduous primary forest' (1995: 358), and that 'before about 1850, most of the present southern sector of the zone consisted of virtually uninhabited virgin high forest owned largely by Akyem people' (1995: 363). For these observations, they cite so many authors that to question their accuracy may seem foolish.[16] But while there clearly was relatively dense vegetation in this region prior to the 1850s, the question arises as to whether it was indeed closed or 'primary'.

Most of the authors cited turn to the comments of Isert, who visited the the Akuapim scarp around 1788. Yet caution is needed in inferring 'primary deciduous forest' from Isert's description. Certainly on the scarp slope itself, Isert did describe 'lofty trees with impenetrable bush underneath cover[ing] the rocks' (1788/1792: 161). He describes Aburi as 'wooded everywhere'. Indeed, he notes that 'Game do not do well because of the lack of grass which cannot thrive in the impenetrable forest.' Yet when he describes huge trees, he qualifies this by saying they were found near water sources (i.e. in humid places, and probably protected): 'Close to the town near a water source, there are trees of an incredible circumference (one 15ft diameter), like a round tower in appearance (1788/1992: 164)'. This leaves one to hypothesise that the character of the forest elsewhere was different. The other trees which he describes are pioneer species, suggesting that the forest was not old growth. Thus he noted a tree with tulip-like flowers which can be identified as *Bombax buonopozense* (see Dalziel 1937: 117), a tree common to secondary forest and farmland. More fundamentally, in this forest he describes yam and plantain (banana) cultivation, writing that: 'The Aquapims use the fruits of the plantain tree . . . a tree bearing fruit the whole year and found throughout the forest (1788/1992: 167)'. Oil palm and raphia palm 'grow wild here in great abundance, but are also cultivated'. It becomes clear that the character of 'the forest' was more like a home garden. While he estimates the population of the region at a mere

9,000 persons, he goes on to say that: 'Their numbers must have been tremendously reduced since towns are mentioned here and there which once must have been flourishing, but of which nothing is known today except the name' (Isert 1788: 172; see also Johnson 1964). Thus it could be argued that much of Akwapim's forest as described by Isert was either managed or derelict home garden. What the vegetation had been prior to these remains unclear.

Early colonial maps of this south-eastern region also distinguish its vegetation from dense forest. Neither Freeman (1898) nor Chipp (1922b) classifies it as within the dense forest zone (Map 4.5). Chipp puts it within the forest–savanna transition zone. More radically, Freeman's (1898) vegetation map describes the southern marginal region as 'hills clothed with thin forest and alluvial plains covered with grass 3–6 feet high', and the western southern marginal as 'thin forest (transition of open bush to dense forest)'.

East of the Volta

A large part of southern Ghana, east of the Volta river, is considered by modern authors to lie within the closed forest zone, and to have been deforested during the twentieth century. Dorm-Adzobu, for instance, suggests that 'Large-scale cultivation of cocoa . . . started in the first two decades of this century when the virgin forests of Central Togoland (mostly in the Buem-Krachi district), were colonized by Ewe farmers' (1974: 45). Once part of the German colony, the region was transferred to British mandate following World War I. In 1926, the Report on Togoland under British Mandate (British Togoland later became part of Ghana) carried a general description of forests in the Ho District (see Gent and Moor 1927). This suggested the presence of deciduous forest in parts of the Tempere Oprana range (i.e. a small southern portion of today's Kabo reserve), on the Western Kunja range (i.e. the western part of Odomi river reserve) and between Kajebi and Wawa (another small portion of Kabo reserve). Apart from these small areas, they found only gallery forest, or what they called 'transition' forest, a mixed humid/savanna forest formation (Gent and Moor 1927). The authors argued that there was an ongoing process of savannisation of forest, and that rainforest had, in the not remote past, clothed the whole country.

Thus while the authors noted the presence of certain forests, they also suggested that savannas were extremely well developed in these districts, and indeed that in the areas south of the 7° parallel, no large forest patches existed. This is confirmed for the 1890s by detailed German maps of the region at 1:100,000. Thus, at least the southerly third of the so-called forest zone east of the Volta was savanna in the 1890s and could have been so long, long before.

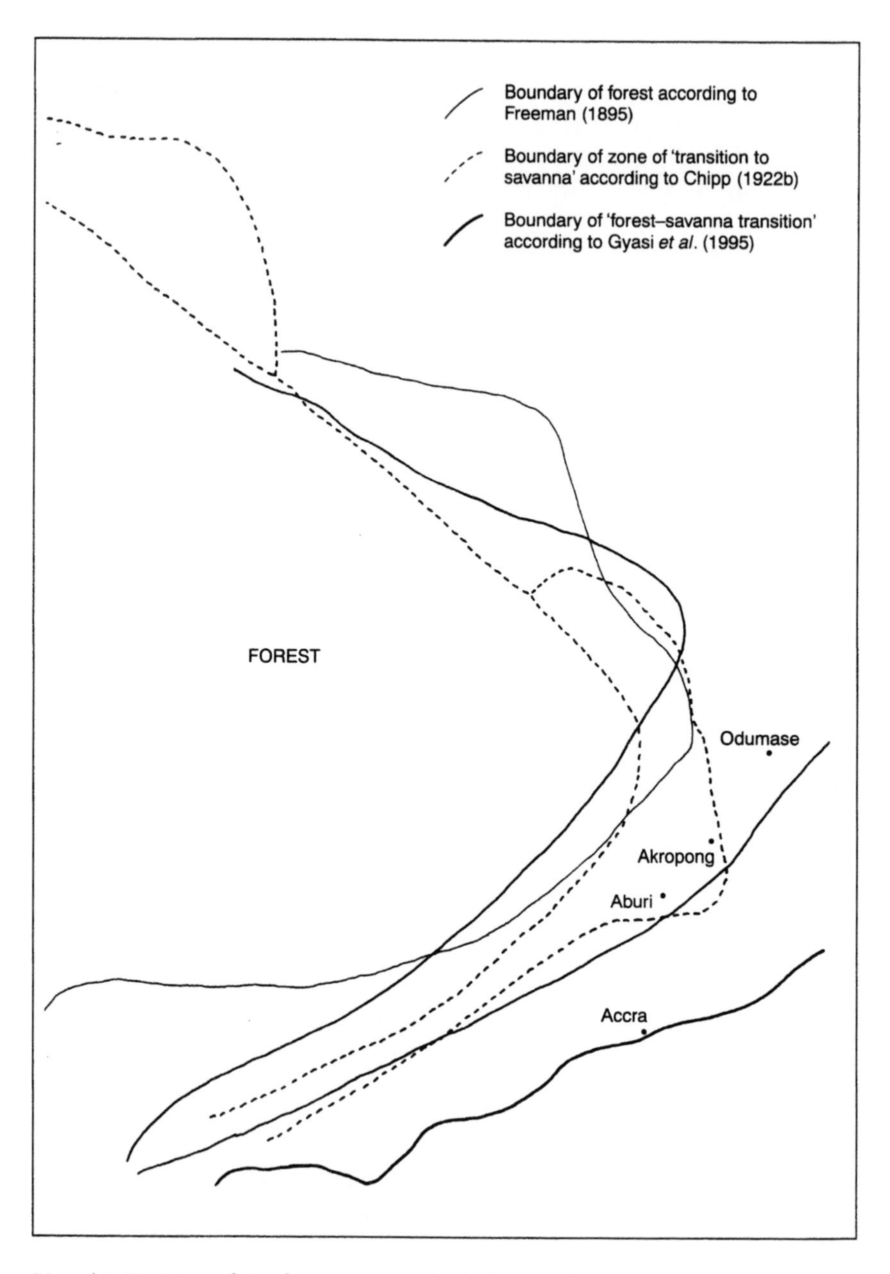

Map 4.5 Position of the forest–savanna boundary in the south-east transition zone in early and modern works.

To the extent that Gent and Moor suggested the presence of extensive transition forest north of the 7° parallel, they quickly and explicitly retracted after spending more time there demarcating potential reserves. Thus in 1930 they suggested that areas which they had originally believed to be sparsely inhabited and largely covered by a good growth of deciduous forest were already under shifting cultivation, consisting of secondary bush. This was so for the Togo plateau and Odomi forest reserves, for instance.[17] This assessment was more in keeping with earlier German reports of the paucity of forest cover on these hills.

When the reserves were created, they amounted to only 66,000 ha, largely on hill ranges and much perhaps not under forest. The only flatland forest in Ho district was the proposed MenuWara reserve. Thus, in the region east of the Volta one might assess the extent of 'forest' – whether 'transition' or dry semi-deciduous – around 1900–20 at little more than 70,000 ha. Indeed, German analysts from 1911 estimated this region, combined with what later became Togo, to have mountain forest cover (excluding gallery forests) of only c. 60,000 ha (Metzger 1911). This can be contrasted with the 500,000 ha of this region which are today considered to lie within the closed forest zone and to have lost forest during the twentieth century. In short, while the incorporation of Togo into the forestry statistics may have increased the area of the 'forest zone', it did not (in reality) much increase the area of forest. What it has increased is the supposed area of twentieth-century deforestation.

Further evidence of forest advance into savanna

The possibility that much of Ghana's forest zone (especially in the dry semi-deciduous and southern marginal zones) was savanna in historical times is tantalisingly suggested by several further data sources. First, the archaeologist Posnansky suggests that savanna inliers as found near Kumasi may well be relics of a past drier climate pre-9000 BC, perhaps maintained by settlement (Posnansky 1982). But Posnansky here overlooks evidence for more recent drier phases. The presence of ancient savannas associated with the Kintampo culture now surrounded by dense forest, such as near the village of Boyase near Kumasi, is suggested by Vansina as a possible indication of forest advance on savanna since 3,000–4,000 years ago (Vansina 1985: 1315; see Anquandah 1982: 6–7). This is certainly consistent with climate history data which show evidence of extreme aridity on the coast and around Lake Bosumtwi at this time (Talbot 1981; Talbot *et al.* 1984; Schwartz 1992; but see Maley 1996). More importantly, if Nicholson (1979) is correct, it is strongly possible that the present phase of rehumidification and forest advance into the savanna zone reflects recovery from more recent dry phases in the forest region from 1300 to 1800.

Second, evidence from termite mounds suggests that areas under forest in the eastern region were, in an earlier period, under more open vegetation. In 1946, Charter noted that scattered throughout Ghana's forest zone are the remains of old termitaria (*Macrotermes* sp.) built, he suggested, when the forest was more open for either climatic or biotic reasons (Charter 1946; see also Jones 1956 in Nigeria).

Third, and intriguingly, there is also oral testimony concerning a savanna past nearer to Kumasi, at the time of the rise of the sixteenth- and seventeenth-century Denkyira state. Ivorian Akan mythology concerning their historic leader, Ano Ansema, suggests that 'he appeared among the Denkyira at Apibweso' and that 'before, at Apibweso, there were no trees, there was nothing, only short grass' (Perrot 1982: 40).

Fourth, Hawthorne argues, on the basis of the contrast between high species diversity in the wet evergreen zone of the extreme south-west and the low species diversity of the moist semi-deciduous forest, that the vegetation in the latter zone may be 'scar tissue, a recently assembled group of mainly widespread, well-dispersed species, covering up after some immense disruption of this area and barely infiltrated by rarer species which could occur there'. He suggests that 'perhaps widespread farming, elephant damage, or fire and drought (e.g. 1500 AD, 3000 BP or 8000 BP) has been responsible' (Hawthorne 1996: 138).

The vegetation history of Ghana's transition zone is clearly extremely complicated. It is certain, however, that a broad-brush analysis showing relentless one-way savannisation does no justice to this complexity. It proves incorrect to assume that large tracts of savanna are 'derived savanna', and that savannas in what is today classified within the closed forest zone, or on its margins, were once forest. Indeed, such views obscure demonstrable instances of forest (or forest fallow) advance over savannas in certain regions, whether due to purposeful enrichment (e.g. in establishing palm forests or cocoa groves in savannas), to climatic rehumidification, to depopulation or to forest reservation.

Conclusions

The extent of the forests which supposedly existed in Ghana in the early part of the twentieth century has been vastly exaggerated. Statistics which give the impression of 8–10 m ha of forest existing in the 1880s are spurious, deriving from a false fit between two forms of calculation: extrapolation backwards of rates of deforestation calculated in more recent periods, and assumptions that what is today defined as the forest zone was indeed covered with forest.

An historical analysis suggests that for the purposes of examining change in forest cover, it would seem more prudent to define today's dry semi-deciduous fire zone as lying outside the forest zone; in other words as savanna, albeit with some patches of forest or transition woodland. This would, indeed, return to the definitions of the forest zone used by early observers. By thus excluding c. 500,000 ha east of the Volta, c. 500,000 ha of land in the northern transition, and the entire southern marginal zone of c. 236,000 ha, the area of the forest zone could be recalculated at c. 7,000,000 ha, which is in keeping with the original assessments made by Chipp (1922b) and the *Gold Coast Handbook* (see Table 4.2). If one estimates, following Chipp, that in 1922 c. 2 m ha were farmland then the amount of forest prior to the cocoa boom may have been at most c. 5–5.5 m ha, of which perhaps 2 m ha remain. In short, we are looking at forest loss during the twentieth century of around 3.5 m ha, perhaps half of the 6–8 m ha suggested by most modern authors.

Moreover, it is highly questionable whether any weight should be given to the c. 1900 baseline. Prior to this time there were at least two centuries of turbulent demographic change, including major episodes of population decline. Large areas of the forest extant in 1900 had regenerated over areas depopulated since the seventeenth century. To present the forest zone as recently populated is to minimise the impact of the slave trade and associated warfare on Ghana. Clearly the extent and importance of past habitation should also be important for understanding the ecology of these forests.

Nor does the distinctiveness of Ghana's forest reserves today, as noted from orbiting satellites, necessarily indicate that 'deforestation has reached the reserve gates'. First, such a pattern could owe its existence to the shape of forest patches when they were reserved in the first place. Second, the opposite process may be more apposite: the forest has grown up to the reserve edge following reserve establishment, protection and fire management. Different explanations probably apply to different regions. Even so, it is probably appropriate to argue (with the exception of the disputed reserves in the Juaboso and Sefwi Wiawso district of western Ghana) that all the forested land that could be reserved in the 1920s and 1930s was reserved. As we have seen when examining the margins of the forest zone, much of the land which was reserved was not, at the time, actually within the closed forest zone, or covered by forest. Comparing the current state of Ghana's forest reserves with their state at the time of reservation, rather than with an imaginary 'natural' climax vegetation state, would seem to provide a more useful picture of the effectiveness or otherwise of Ghana's conservation policy. Yet it might not prove acceptable to conservationists, carrying less rhetorical force of crisis and decay.

Notes

1 We are indebted to Dr William Hawthorne for his remarks and corrective comments on an earlier draft of this chapter, a working paper entitled 'Conservation and the new forests of Ghana'.
2 This is the zone usually associated with Hall's loose definition of forest as

'vegetation dominated by trees, without a grassy or weedy under-storey, and which has not recently been farmed' (Hall 1987: 33).

3 Estimations of present forest cover generally fall between 1,500,000 and 2,000,000 ha. At the upper end, de Lange and Neuteboom. (1992) suggest a figure of 2,100,000 ha of remaining forest, whereas at the lower end, EPC (1991) and Paivinen and Witt (1988, incorporated into Sayer *et al.* 1992) estimate 1,700,000 ha. Such discrepancies seem to hinge on variable estimations of the forest area outside the reserves:

Source	*Area (ha)*
de Lange 1992	500,000
World Bank 1988	270,000
Sayer *et al.* 1992	100,000

4 Zon and Sparhawk's figure for Ghana relied upon early estimates of forest cover made in Ghana by McCleod (1920) and Unwin (1920).

5 In an earlier preliminary publication concerning the work (Fairhead and Leach 1996b), we were perhaps not sufficiently cautious in our use of Meniaud's work.

6 Of which 1,528,243 ha was reserved and 1,970,000 ha unreserved forested land.

7 Boateng (1962) estimates that cocoa farms came to occupy 47 per cent of farmed area by then.

8 Cocoa seems to have altered the nature of fallows, being established in fields where cocoa was intercropped with food crops for the first few years, and selected trees were permitted to regrow to provide a broken canopy for shade. In many regions, cocoa was generally established in fields and came to dominate land which would otherwise have become fallow.

9 Johnson (1981) considers Wilks' model to be unrealistic for several other reasons: (a) if lands were cleared in the sixteenth century, she suspects that they would not have remained permanently cleared; (b) imported labour was used for porterage; (c) Kumasi may have been intensively occupied prior to the seventeenth century. She argues that land was occupied, then released; the inhabitants did not find an equilibrium with their environment.

10 A casual reading of Hawthorne and Musah would suggest that 50 per cent of Ghana's 1,774,500 ha of forest reserves are 'mostly degraded' or worse (1995: 15). Yet of the reserves classified in this way, it could be argued that more than half were 'mostly degraded or worse' when originally reserved. Thus of those reserves classified as having 'no forest', 131,000 ha belong to the Juaboso district complex of reserves, which are exceptional as they have been recently reserved against strong local opposition, and the rest to the southern marginal or dry, semi-deciduous fire zone, which cannot be considered to have been forest in the first place. Of the area classified as 'very poor' in forest, that in the moist evergreen falls into the same category as Juaboso, being recently reserved; and that in the moist semi-deciduous south-east and the dry, semi-deciduous region (131,600 ha) was either not 'forest', or at least not in good condition when reserved. Thus at present we can assume that only Bia North (35,600 ha) and Desiri Goaso – which was reserved late in 1954 (15,100 ha) – present reserves seriously degraded since reservation. Of the forest classified as mostly degraded, that in the dry, semi-deciduous zone might be considered to have been in this state (or worse) when reserved.

11 Information from inhabitants of Degedege during field visit by the authors, June 1996.

12 Translated in *Ghana Notes and Queries* NQ: 16.

13 This vegetation is subject to occasional ground fires in the litter layer of the forest floor. The effect of ground fires is to kill many of the smaller shrubs and young trees, but the fire is seldom sufficiently intense to kill large trees; the forest thus survives (Swaine *et al.* 1976).

14 This, Moor asserts, 'was the first known fire of any appreciable scale in this locality and burnt out a semi-circular belt a mile wide on a four mile radius' (1936: 3).

15 Yensiso [in Akuapem, Yensiso, Adenya, Gyamfiase, Kokormu], Amanse [south Akyem, Amanase, Whanabenya] and Sekesua [Krobo, with Osonson].

16 Notably Dickson (1971), Gold Coast Survey Department (1949), Hall and Swaine (1976), Hill (1956), Johnson (1964), Kwamena Poh (1973) and Taylor (1960).

17 Report on *Togoland Under British Mandate for the Year 1929*, London, HMSO (1930: chapter XXI, Forestry).

5

BENIN

Introduction

Benin's forest history, as presented in the national and international literature, seems to have been one of inexorable forest loss. Evidence of extensive forest cover in historical times – dated by some to c. 1900 – and for its subsequent loss has been based both on certain early descriptions of the region and on deductions concerning past vegetation cover and past climate. In this chapter we review some of the evidence concerning forest cover around 1900. As we go on to examine, however, the Benin case is prime among West African countries for exemplifying how forest history has come to rely on deductions from a set of vegetation 'indicators'. While these have been introduced in earlier chapters, this chapter focuses on them in some detail. We argue that for each of the vegetation patterns which have been interpreted as the product of relentless and almost complete forest decline, an alternative interpretation is not merely possible, but probable. As analysis of Benin has relied so heavily on this type of deduction, by questioning the underlying reasoning the chapter opens the way for a radical reappraisal of Benin's vegetation history and the role of people in it. It raises the strong possibility that forest history, as usually represented in Benin, has obscured skilful land management techniques which have been used over many centuries to enrich landscapes and vegetation. (See Map 5.1.)

Forest loss

In their influential article, Gornitz and NASA (1985) argue that Benin had large tracts of forest around 1900. They suggest that 100 years ago, the 50-mile-wide coastal plain housed around 1.12 million ha of dense, tropical forest, going on to summarise that:

> Approximately 500,000 ha of equatorial forest were reported around 1930 (Ann. stat. A. O. F., 1933/34) although according to Aubréville (1937) the coastal forest of Benin had nearly all

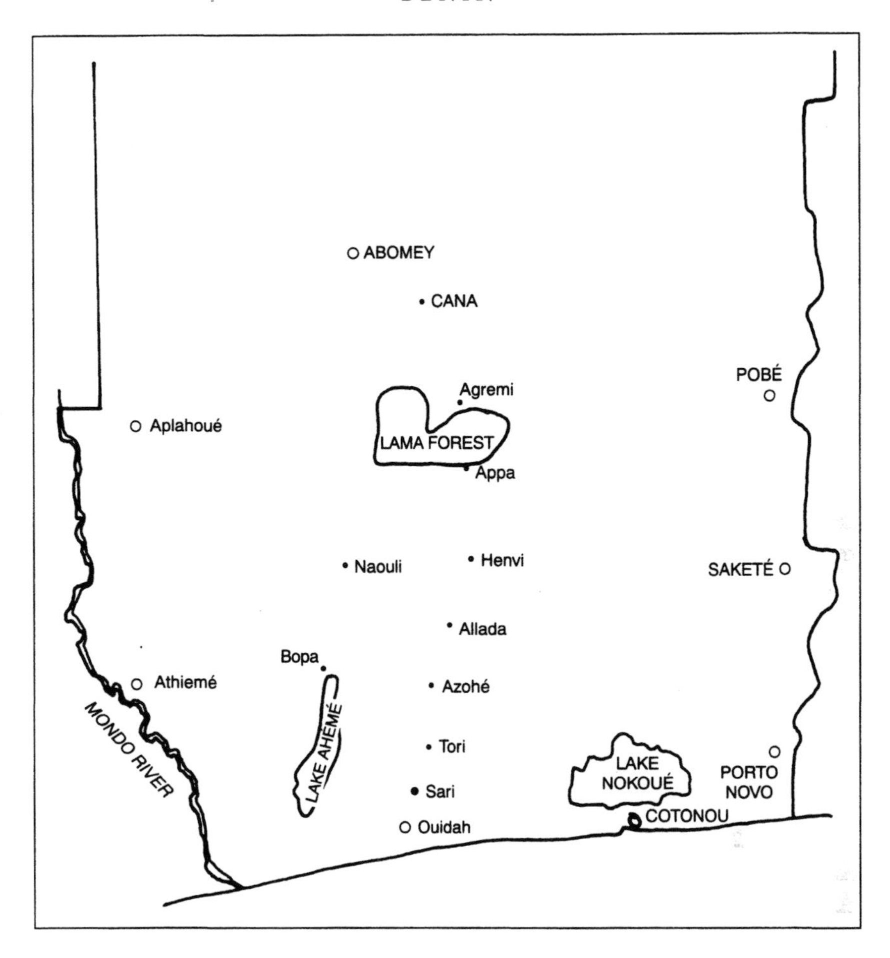

Map 5.1 Regions of south Benin described in the text.

> disappeared by 1937. By the 1970s, Persson (1974) reports 200,000 ha of closed forest (1971) and FAO (1981) only 47,000 ha – a large difference, yet clearly much reduced from estimates of 50–60 years ago. As much as $\frac{3}{4}$ of the original woody vegetation is estimated to have been either cleared or converted to secondary formations or bare soil (FAO 1981). The present deforestation rate is 3.2%/year.
>
> (Gornitz and NASA 1985: 316)

Most authors have been more cautious about dating the presence of intact forest, but the image of relentless decline is reproduced in all recent works which review Benin's forest history. The FAO (1993) assessment, for example, suggests that 56,700 ha of moist deciduous forest are being lost each year. The FAO 1980 assessment states boldly that 'the impoverishment of woody vegetation, essentially due to the pressure of itinerant farming, overgrazing and bush fire, continues in an inexorable manner' (FAO 1981: 23). Sayer *et al.* (1992) suggest that the predominant vegetation of Benin was originally dense semi-evergreen or deciduous forest, with closed-canopy tropical moist forest once covering 1.7 m ha, and now covering only 42,000 ha.

Most of the authors who consider coastal Benin to have been forest around 1900 have based their analysis on the works of certain early geographers. Gornitz and NASA, for example, drew on the global forest assessment made in 1923 by Zon and Sparhawk, who suggested that 'after the first mile or two of sandy waste, the 50 mile wide coastal plain is for the most part covered with dense tropical forest' (1923: 861). While they do not cite their sources clearly, the picture they present was in keeping with that of the French geographer Breschin, who in 1902 estimated the forest area to be c. 1,200,000 ha (see Map 5.2). Breschin based his analysis on the sparse travel accounts of Albeca (e.g. 1894) and Plè (1900). Taken at face value, these travel accounts support the idea of large tracts of coastal forest. Plè's mission delimiting Benin's eastern frontier noted that from Dopetu near the coast, up to 35 km inland (to c. 6.55° N), the 'lightly undulating soil is covered with a rich vegetation. It is the virgin forest in all its splendour, as well as all its difficulties of penetration' (Plè 1900: 2).

In the more central region, several travel accounts detail what seems to be quite extensive forest cover. Some of this is reviewed later in the chapter, when examining historical accounts of the road from the coast to Abomey. As a summation, Albeca writes that:

> Beyond the lagoons, as far as [the marsh of] Lama extend vast forests; all the tallest trees, all the tropical species are found there, ficus, cailcedrats, silk cottons [*Ceiba pentandra*], mahogany [*Khaya* spp.], docotchou or black wood, the iroko [*Milicia excelsa*] or Jebe, a red wood that the inhabitants saw into planks of 2–3 metres and use in the construction of houses in Porto-Novo, Ouidah and Agoue. Everywhere an impenetrable bush interlaced with lianas, and crossed by paths leading to the villages. From time to time one sees clearings and cultivated fields.
>
> (Albeca 1894: 170)

Yet we argue on the basis of other evidence from the same period and earlier that the image of abundant vegetation which is portrayed provides little basis for assessing subsequent changes in forest cover.

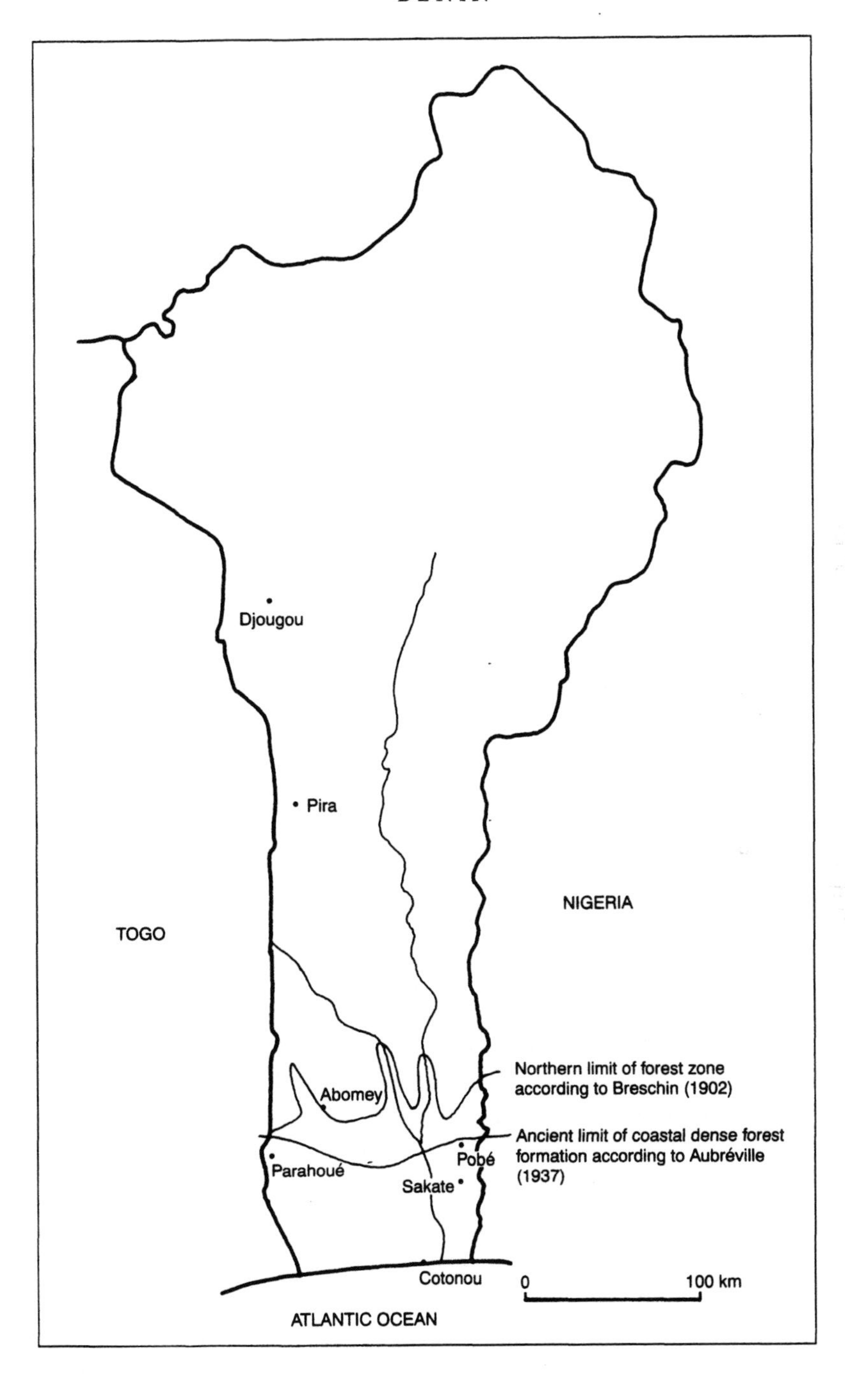

Map 5.2 Benin and the forest zone according to Breschin (1902) and Aubréville (1937).

Alternative perspectives: early maps and palm groves

First, and fortunately, a map exists detailing vegetation for what was Dahomey, dating from 1893 (Map 5.3). While there is certainly room for misinterpretation, dense vegetation can nevertheless be distinguished from open farmbush or grassland at a first approximation. Furthermore, this map can, in effect, be ground-truthed contemporaneously by the accounts of the travels of early botanists to the region. For example, in the east, from Porto Novo to Saketé, Chevalier (1910, 1912b) noted what the map shows: a prosperous plain covered with farms (which Chevalier described as well-tended maize and yam fields), and clumps of dense vegetation, described as palm groves, interrupted by scattered groups of houses. 'At Saketé [the northern edge of the old map] the fields become sparser, and instead vast fallows appear covered with impenetrable bushes mixed with softwood trees (1912b: 34).' This vegetation Chevalier distinguished from the islands of dense forest found in the area. It is immediately apparent that the eastern frontier region which Plè visited was exceptional, as were certain parts of the journey between the coast and Lama described by Albeca. The absence of forest at the turn of the century in some of the areas not covered by this map can also be evidenced in western areas of Benin. In a 1909 report on forests in the large western administrative region of Aplahoué, for example, an area within early definitions of the forest zone, a mere 1,600 ha of forest were noted, excepting that on the floodplain of the Mono river (in Wartena 1992, annex D).

Second, most of what was called forest by the early French explorers and administrators on whose reports early accounts are based, was not 'forest' as conventionally understood. Rather it consisted of vast palm groves, albeit with large parts seemingly derelict. Thus, the botanist Chevalier, who spent six months examining the vegetation of Dahomey, stated in 1912 that:

> Oil palm is the principal wealth of the colony. The densest stands are found between the coast and the southern border of the wooded and marshy region of Lama [i.e. the area noted by Albeca, and the major block of forest marked on the 1893 map] on a band 60 km wide and 110 km long, going from the frontier of Togo to that of Lagos; it is no exaggeration to say that within these limits, the palm covers all the land and gives almost everywhere its aspect to the landscape. In some places, the uncultivated bush seems to dominate, but when one examines it closely, one finds that this bush covers land in fallow, and among the trees and bushes, live in close ranks *Elaeis* [oil palm] crowns, obscured by the wild vegetation. They are almost unproductive, but it is sufficient to clear around them and take off the dead leaves to make profitable plants.
>
> (Chevalier 1912b: 30)

Map 5.3 Early vegetation cover in southern Benin in 1893 derived from a 1:100,000 map (Archives Nationales de la France, Aix-en-Provence, AF43).

Chevalier is clearly at pains to contradict what he considered to be misrepresentations of Benin's vegetation in early accounts, suggesting how to the untrained eye it would be easy to mistake derelict-looking farmbush and oil palm for the 'majestic forest'. Furthermore, even though Albeca wrote about 'vast forests', his own map (1894) betrays this vision, as virtually all the 'forest' is described as forest of palm trees, or merely as palms. Other early maps are equally explicit. The national maps of the 1920s (1:1000,000), for instance, clearly identify dense vegetation as palm forest.

Many analysts of West African vegetation have considered the presence of oil palm trees to indicate the past presence of high forest, now lost (e.g. Keay 1959b; Allison 1962). But this was not the conclusion of those who observed Benin's oil palm forests. Chevalier and later Aubréville argued that much of Benin's palm forest had been established in savanna.

> The large extension which the inhabitants give, for some time, to the cultivation of crops, especially maize, leads to the development of palm groves, for in clearing, the inhabitants conserve the palms and start to look after them from the day that the soil is first used. It is no exaggeration to affirm that in Dahomey, all fields of maize made from the savanna/bush and sometimes also, unfortunately, from the forest (which has thus, and little by little, disappeared almost completely) become subsequently a palm plantation. One can thus hope that in a few years, the vast territory of 600,000–700,000 ha defined above could become an immense plantation of *Elaeis*.
>
> (Chevalier 1912b: 30)

Aubréville claimed to be able to distinguish (from species composition and biodiversity) those palm groves that had been established in forest from those which had been established in savanna. Palm groves very recently established in savanna could easily be detected, for example, if they contained valued savanna trees such as *Parkia biglobosa*. The absence of these trees and the presence of forest flora indicated to Aubréville where palms had been established in forest. Nevertheless, it is less certain that these methods would be capable of identifying in what vegetation much older palm groves had been established; and many of Benin's palm groves are ancient, dating from before 1700. In this region, closed-canopy formations such as dense palm groves very rapidly become diverse and acquire flora associated with forest, especially when left 'semi-derelict'. Savanna species disappear and forest species can easily colonise from patches of forest in the neighbourhood, whether village 'sacred' forests, streamside riparian forest, or other 'forested palm grove'. Thus finding a palm grove which contains forest species and no savanna ones may be no indication that it was established 'in forest'.

Reconsidering deforestation

While appreciating the establishment of palm forest in savanna, early scientists clearly did consider large tracts of forest to have been deforested. But it is important not to confuse what observers such as Chevalier thought about the past with what they saw at the time of their visits.

It is clear that Benin's forest zone was not covered by high forest when Chevalier visited at the beginning of the twentieth century. By then, as Chevalier put it, the forest had already disappeared 'almost completely'. The only dense forests which Chevalier saw were very small patches in some swamps and on sacred sites, which he considered to be relics of the past vegetation. Equally, he observed 'relic' trees which he supposed once to have been surrounded by forest because they had the tall, straight trunks characteristic of a forest form. It was in the eastern section of the so-called 'forest' zone that Chevalier noted the presence of many forest islands:

> From Saketé to Pobé one crosses swamps not used for crops, and covered with a high forest canopy in which one finds most of the beautiful species of Côte d'Ivoire. These islands are, however, of very small extent, but from the sight of some giant trees distributed across the fields and fallows, one quickly acquires the conviction that the large virgin forest has, once, covered all the land.
>
> (Chevalier 1910: 428)

Chevalier specified the trees as *Chlorophora excelsa*, *Antiaris africana*, *Alstonia*, *Triplochiton* and *Irvingia* spp. Chevalier's findings on his 1909 tour were seen again by Aubréville in 1937. Speaking of the coastal forest regions, he stated that:

> [O]ne cannot speak of these [forest] formations in the present. Their almost total disappearance is a deed done a long time ago. . . . It is hard to believe today that all this coast region was once covered in thick forests. This is now no more than palm forest and crops. As testimony, however, there remain the fetish woods, and some forests on the way to rapid destruction on the frontier with Nigeria, near Pobé . . . a few isolated trees in the fields, somtimes a clump of ancient trees, a sacred grove, in the proximity of a village.
>
> (Aubréville 1937: 42)

Merely from comparing the observations of Chevalier and Aubréville, and noting contradictions between them, we can begin to appreciate major problems in their deductions concerning forest past. Aubréville travelled on the same route as Chevalier between Saketé and Pobé. His observations of isolated trees and forest islands led him to conclude that the land had still been forested 10 years before (i.e. around 1927). In 1937, he wrote, a road was opened to from Saketé to Pobé 'across the forest' and that this 'was the signal for the massacre. All along this route, from Saketé to Pobé, the destruction was at full tilt. Today, large isolated trees or forest islands are still found in the fields. Evidently, the felling only dates from a few years' (Aubréville 1937: 44). Yet as we have seen, Chevalier (1910, 1912b) described a very similar vegetation along this same route, but some 30 years earlier, falsifying Aubréville's interpretation of recent forest loss. It was not 'a forested region' prior to the road being built and there was, it seems, no 'massacre'. When Chevalier saw this landscape he also deduced that it was the product of recent deforestation. But in this, was he in fact making the same mistake as Aubréville?

Similar observations of landscape and deductions about its recent past have been made by many subsequent visitors to the Saketé–Pobé. The administrator Tereau noted in the 1930s, for instance, that: 'The plateau [of Pobé–Saketé] is covered in luxuriant vegetation and magnificent palm groves, which get more open the more one moves towards Pobé.' The plateau was 'once covered by the large forest', and the seasonally inundated zone has 'here and there some vestiges of the equatorial forest, and large scattered trees of *Ceiba pentandra* and *Adansonia digitata* [baobab]' (Tereau and Huttel 1949: 62).

The possibility arises that both Chevalier and Aubréville, and the many subsequent analysts who have been influenced by them, have been badly misreading Benin's landscape history. The major landscape features which led them to suspect deforestation are those which dominate our case for re-evaluating forest change throughout West Africa: (a) the idea that trees in their forest form grew in forests; (b) the idea that islands of forest are relics of a larger expanse of forest; and (c) the idea that bush fallow is degraded forest (see Map 5.4).

Isolated trees with their forest form

Both Aubréville and Chevalier noted the presence of tall, straight-boled trees of 'forest species' growing in fields, considering them to have taken this 'forest form' because they grew up in forest. Yet the limited number of tree species which they cite are all known to be capable of growing into this shape whether in the open or in forest. Furthermore, these trees are very fast-growing in fallowed land. When fallows are farmed again, large individuals are frequently preserved in new fields for their valuable timber or other products. Thus, while the presence of these tall forest trees towering as individuals over an otherwise farmed landscape easily gives the impression

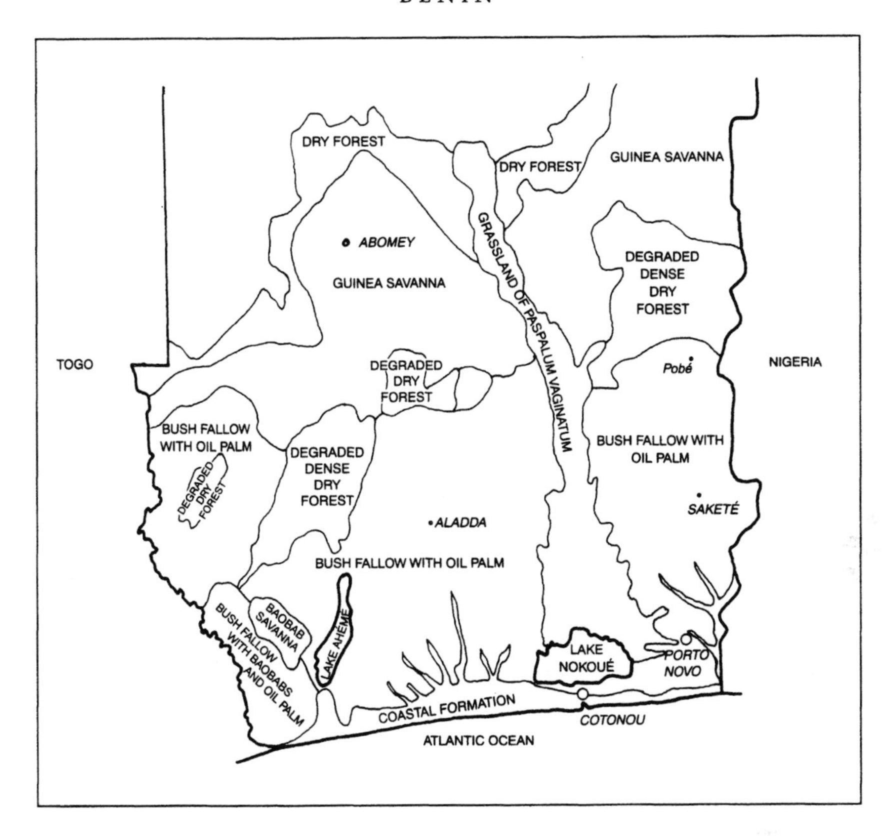

Map 5.4 Location of major vegetation features in southern Benin (Mondjannagni 1969).

of recent degradation, more local historical evidence would be required before accepting this.

It is therefore quite possible that the 'isolated trees' which early travellers saw grew in farmers' fields and have not grown in a forest of which there is no other sign. It is worth noting that even today most of the timber which reaches Benin's sawmills is of *Milicia excelsa* or *Afzelia africana*. Most of this timber derives from trees preserved on farmland, along roadsides and shading tree crop plantations (see Sayer *et al.* 1992). These trees, usually less than 60 years old when felled, cannot have grown in forest, for, as we have seen, forest did not exist 60 years ago.

Forest islands

All botanists, foresters and agronomists in Benin since the time of Chevalier have treated the forest islands found there as relics. Dense forest is represented only by these forest islands, as has been the case since Chevalier's day.

Such forest islands have received a great deal of botanical attention because apparently, as Aubréville argued (1937: 43), with the aid of these forest patches and isolated trees, 'from all these spared forest elements, one can reconstitute in spirit what the original autochthonous forest was like'. On the basis of this analysis, Aubréville felt able to demarcate where the forest had once existed: 'The northern limit of the ancient coastal forest was bordered by Pobé and Parahoué. This forest band had an average width of 60 km in Benin. It extended near Parahoué 80 km from the coast' (Aubréville 1937: 53).

This position that forest islands are relics also dominates more recent studies of Benin's vegetation. Adjanohoun notes that 'the forest exists only as relics. These are either sacred forests, or those protected in certain agricultural research stations (Pobé, Naouli)' (Adjanohoun 1966: 32). Mondjannagni (1969) considers forest islands to be 'dense, humid, semi-deciduous rainforest, veritable relics of an ancient continuous forest cover'. Gu-Konu (1983: 32, cited in Gayibor 1986) considers these forest patches to be degraded relics of a climatic climax forest, suggested by the stands of *Antiaris africana*, *Milicia excelsa* and *Triplochiton scleroxylon* scattered here and there in the landscape (see also FAO 1981: 25).

Many of these forest islands either encircle villages, or are fully or partially preserved on the sites of abandoned villages. As Wartena notes, at the beginning of the century, most Adja villages (south-west Benin) were surrounded by a band of forest: 'this band protected villages against enemies, served . . . as a source of forest products, as a wastebin and as a toilet and contained sacred places' (Wartena 1994: 77). Since 1920, many of these forest islands have been converted into fertile farmland, with the exception of small parts maintained for ritual purposes. Brouwers makes much the same point: 'Originally, each village had been surrounded by an intact circular zone of original vegetation. This zone is still to be seen in some villages in 1957. The zone used to serve in former days as a defensive barrier to hide from slave traders' (Brouwers 1993: 84). For a number of authors focusing on the relationship between land management and vegetation cover, then, peri-village forest islands provide an insight into the region's 'original' vegetation. That forest islands represent the original vegetation has also been incorporated more deeply into agricultural research, where forest island soils are used as 'baselines' or as 'control' in studies which consider the potential or 'original' fertility of its soils (see, for example, Djegui 1995). Fertility under other land uses is compared with this 'original' fertility.

But are these deductions valid? Are forest island soils and vegetation 'representative' in the way they are assumed to be? There are several reasons why this is very probably not the case.

First, because these forests are generally found either around existing villages, on the edge of villages, or over ruined village sites they cannot be taken as representative of vegetation in the rest of the landscape. They are generally growing over anthropogenically transformed soils. As Brouwers notes: 'Where the forest belt [around villages] is still intact, household refuse, human excrements (the forest is used as a latrine) and droppings from small ruminants and poultry enrich its soil' (1993: 92). As Keay (1947) and Thomas (1942) pointed out long ago for Nigeria, such super-fertile 'anthropogenic soils' may support a dense vegetation where, prior to habitation, they could not (see Fairhead and Leach 1996a). So can the soils of these islands be considered as representing 'natural' fertility? It is in error that the reserved forest patch now in Pobé agricultural station is used to define 'original soils and vegetation' as it was once part of the village forest island of Pobé (Aubréville 1937: 44). As many archaeologists have shown, the effect of such enrichment on soil nutrient levels and pH, as well as on soil macro and micro flora, can endure for many centuries if not millennia. Thus, the forests also preserved over ruined villages – often housing the most powerful sacred groves – can hardly be treated as natural. Indeed, to consider these anthropogenically enriched soils as natural may be to over-estimate the nutrient accumulation under natural forest and, in comparison, to render disproportionately small the soil-regenerative effects of other vegetation cover.

Second, as is argued elsewhere in this volume, peri-village forest islands and forests abutting villages are frequently the product of settlement and active forest management (and tree planting) by inhabitants, with known cases from Sierra Leone, Guinea, Ghana and Gabon. They may be created to serve as fortresses, to protect villages from bush fire and to create a source of privacy and forest products. To date, this possibility has never been considered by those examining the nature and origin of Benin's forest islands. In the rare cases where forest islands are not associated with villages or ruined villages, they may still have been managed in very particular ways to render them sacred, including the planting of particular trees and the introduction of particular termite species which would favour soil transformation and particular tree species development (see Iroko 1982; de Surgey 1988, 1994).

The potential pertinence of this criticism can be gauged from the choice of forest islands used to comprehend Benin's vegetation and from the botanical composition of those forest islands. Table 5.1 notes those forest islands covered by the two major studies detailing Benin's vegetation, by Aubréville (1937) and by Mondjannagni (1969); the forests which thus provide the data used to reconstruct Benin's original vegetation. As one can see, of thirteen forests studied, only one is not known to be exceptional (that of Toffo, for which no site detail is offered). The others are noted by Aubréville and Mondjannagni to be either (a) village forest islands; (b) covering a ruined settlement site; (c) fetish groves (perhaps also falling into category a or b); or (d) streamside forests (perhaps falling into the other categories too).

Table 5.1 Forest islands used in the study of Benin's 'original vegetation'

Forest island	*Authors*	*Type of forest*	*Notes*
Djougou	Aubréville	Village forest island	
Banté	Aubréville	Village forest island	
Pira	Aubréville	Village forest island	
Bouyierou	Aubréville	Village forest island	
Pobé	Aubréville, Mondjannagni	Village forest island	Gallery forest
Allada	Aubréville	Ruined settlement	
Saketé	Aubréville	Fetish grove	
Naouli	Aubréville, Mondjannagni	[not given]	Gallery forest
Toffo	Aubréville	[not given]	
Foncomé	Aubréville	Village forest island and fetish grove	
Attogon	Mondjannagni	Ruined settlement	No gallery forest
Dogla	Mondjannagni	Fetish grove	[created]
Houézounmé	Mondjannagni	Fetish grove	

It cannot be assumed that the botanical composition of these forest islands reflects 'natural' or 'original' forest. Trees central to both ritual and everyday life are known to be planted or transplanted in this region. While it is often assumed that many tree species acquire use value 'because they are there', the inverse is often the case: they are there because they are important. Certain tree species are known for their importance in sacred performance and what is colloquially known as 'fetishism'; others for their economic and social uses. Thus, for example, the *Triplochiton scleroxylon* which Mondjannagni used as an indicator tree to 'determine the past extent of the rainforest' is, as he also points out, the Oro fetish tree of Yoruba, and as such may be planted, or may appear on soils which have been fundamentally altered by adept elders. *Ceiba pentandra* is not only a tree used for sacred purposes throughout West Africa, but also grows rapidly from the cuttings used very commonly both in fencing and in fortification. *Antiaris africana* (false Iroko) is both planted to install sacred altars in the region (see also de Surgey 1988: 320–34) and used for its bark, which can be cut and beaten into traditional cloth. *Milicia excelsa* is also a fetish tree, planted in sacred sites, in front of houses, and as a 'Palaver tree' (see also de Surgey 1988: 320–34). It is known to thrive on soils relatively rich in calcium, and part of its sacred quality may derive from its links with the presence of past settlements, and farm campments which have become calcium-rich. *Cynometra megalophylla* is known to be transplanted from riparian forest (Mondjannagni 1969). *Ficus* species are often planted both for fruit and for social purposes (Isert, trans. Gayibor 1986: 245). Several other trees are planted around sacred grove altars such as *Newbouldia levis*, *Spondias mombin* and *Dracaena arborea* (Mondjannagni 1969). *Allanblackia floribunda*, *Carapa procera* and *Pentaclethra macrophylla* used to be transplanted for their useful oil grains, *Blighia sapida* and *Chrysophyllum africanum* are planted for their fruit, as is *Daniella oblonga* (Chevalier 1910, 1912b; Dalziel 1937). In neighbouring parts of Nigeria, *Khaya grandifoliola* and *Canarium schweinfurthii* are noted as generally being planted (Lamb 1942).

If one examines the botanical composition of the Benin forest islands as described by Chevalier (1910, 1912b), Aubréville (1937) and Mondjannagni (1969) one finds that many are not highly species diverse and that the predominant trees are those known to be transplanted.

That forest islands in this region might have been the product of people's vegetation management, and enriched with tree species, was in fact noted by much earlier travellers to the region. The most notable is Isert, who writes in a letter of 28 March 1785:

> The area around Fida [Ouida, south-west Benin] is one of the most attractive of all the places where Europeans have settled in Guinea. The ground is level and blessed with meadows in which there are fresh water sources scattered all around. . . . The farther I come into the Bight of Benin, the more enthusiastic I find the people in their worship of idols. At Orsu [Osu], around Christiansburg [now in Ghana], they have no public fetish temples; here [around Fida] they have more than thirty. I have seen some here which have several forecourts and a number of rooms, and are surrounded by beautiful trees. I like to go to such places because I always find those trees there which are rare in the country and are planted ['*places à dessein*' in Gayibor's 1989 French translation] because of their rarity.[1]
>
> (1788/1992: 104–5)

Several of the forest islands featured in Table 5.2 lie in more northern savannas. All these forest islands conceal villages. In 1937, Aubréville considered from analysis of their species composition that these forests were relics of a past wetter climate, protected from fire by their wetter soil and from clearance by their occupation. The drier soils around them, he suggested, might have been forest under past wetter climates, but then were transformed into dry forest – which itself was degraded to savanna by fire. He suggested then that the occupation of these forests was ancient, that their occupation and use as fortresses had saved them from destruction, and that they were now 'condemned' by the peace brought by the colonial regime. Nevertheless, such forest islands in the savannas have a very similar flora to the particular anthropogenic *kurmi* forest islands described in Nigeria by Lamb (1942), the anthropogenic forests which came to cover old Oyo in Nigeria (Keay 1947).

Table 5.2 Tree species in forest islands in Benin

Known subject of anthropogenic enrichment	*Tree species*	*Observed by*									
		Aubréville 1937			*Mondjannagni 1969*					*Chevalier 1912*	
	Name of forest island (y=species observed)	*Pobé*	*Saketé*	*Naouli*	*Foncomé*	*Attogon*	*Dogla*	*Houzo-éunmé*	*Pobé*	*Saketé*	*Ketou*
	Terminalia superba	y			y	y			y		
y	*Piptadeniastrum africanum*					y			y		
y	*Ceiba pentandra*	y	y	y	y	y	y	y	y	y	y
y	*Milicia excelsa*	y			y	y	y				y
y	*Triplochiton scleroxylon*	y	y		y	y		y	y.	y	y
y	*Antiaris africana*	y	y	y	y	y	y		y	y	y
	Celtis mildbraedii					y	y				
	Celtis zenkeri	y	y	y	y	y			y	y	y
	Celtis soyauxii		y							y	
	Celtis pranthii	y									y
	Dialium guineense	y									y
y	*Chrysophyllum albidum*	y						y			
y	*Cola cordifolia*	y		y	y				y		
y	*Afzelia africana*	y			y						y
	Trichilia prieuriana	y				y					
	Holarrhena africana	y									y
	Parkia bicolor	y		y							
	Cordia senegalensis		y							y	
	Cistanthera papverifera		y							y	
	Firmiana barteri				y					y	y
	Macaranga barteri			y					y		
y	*Dracaena arborea*					y	y				

That certain of the forest relics in Benin are recently established was actually noted by Mondjannagni (1969), who argues that the forest island of Houézounmé, cited above, is incontestably older than Dogola; a 'veritable primary forest which has always existed'. But Dogola, he states explicitly, 'is not a primary relic, but a more recent formation (1969: 90).

The interpretation of place names has been used to reinforce the idea that forest islands are relics. Aubréville, for example, supports his deduction of extensive forest loss by arguing that the name of the village Irocogny (meaning 'place of the iroko', *Milicia excelsa*), in the savanna regions, indicates the past abundance of iroko (hence past forest) on the route from Ketou to Idigny (1937: 47). Again such deductions stretch the imagination. In the region, such a name could equally refer to the past presence of one particular large or otherwise special iroko tree, a tree which we have noted is fundamental to religious and social practices in the region, and is frequently planted and transplanted.

While patches of forest have been used to indicate the past existence of much larger forest cover and the erosion of 'original forest' under land use, an alternative hypothesis needs to be considered: namely that forest islands may indicate the capacity of people to enrich and manage soils and encourage forest vegetation to grow on areas which may earlier have been savannas.

Strangely, the forest island issue lies at the heart of recent assessments of forest cover change in Benin. In the FAO 1980 forest assessment, the analysis for Benin was informed by a study which compared air photographs and satellite images of assorted dates for Benin, Togo and Cameroon, conducted by FAO/UNEP/GRID (1980). It was envisaged in that study that site sampling procedures would make the study generalisable to the country, but in fact the researchers ended up, after preliminary investigation, preselecting areas where significant forest cover changes had occurred and which were visible on available air photography. In Benin they selected only one site which had any semi-deciduous forest, and this was one of the forest islands far to the north of the coastal forest zone. The semi-deciduous forest chosen for study was the village forest island of Djougou, which covered 306 ha in 1949. Long ago, Aubréville had noted this particular forest island as 'the most menaced' of the forest islands to be found around villages quite as far north as this in Benin's forest zone (1937: 28). The lower vegetation had even then been completely cut or burnt, and Aubréville thought it doomed. Nevertheless, in 1975 the FAO/UNEP study found it to remain, albeit perhaps 85 ha smaller (or 20 ha smaller and 65 ha degraded).[2]

The quality and dynamics of vegetation in and around peri-village forest islands are clearly no basis for understanding national changes in forest cover. But astonishingly, and on the basis of this one forest island shrinking by some 85 ha between 1949 and 1975, FAO (1981) claimed that 'the results of comparative air photography served as the basis for elaborating the annual rate of deforestation' for all of Benin. This rate had been 1500 ha/year during 1976–80, and was predicted to fall to 1,200 ha/year during 1980–5. That they draw such conclusions on the basis of one, preselected, smaller forest island defies belief for many reasons, apart from the issues of the representativeness or otherwise of forest islands. Suffice it to say here, however, that as the study preselected areas of loss, any gain in forest elsewhere which might counterbalance the statistics would not have been identified.

Fallow vegetation

An analogous interpretive problem to that of forest islands surrounds a rather odd 'forest formation' characteristic of several of the coastal plateau lands (the Comé plateau, a part of the Adja plateau and the Allada plateau), but also found elsewhere. In essence, this comprises baobabs (*Adansonia digitata*) within secondary forest thicket.

In his 1937 work, Aubréville expressed astonishment at seeing baobabs in closed thicket formations in this part of the coastal 'forest zone'. He suggested that this was a degraded form of an original dense semi-deciduous forest type specific to Benin, of which baobabs were a characteristic species (1937: 49–53). His opinion concerning this was crystallised, he said, when he examined the sacred grove of Foncomé (a village between Athiemé and Parahoué). Here he found baobabs in a 'high equatorial forest', and deduced, first, that in this region baobabs exist in equatorial forest; second, that baobabs-in-thicket are a degraded form of this; and third, that baobabs present in the savannas are relics of this. Such historical deductions were in keeping with his pejorative views of African land management and convictions of ongoing savannisation.

But as Mondjannagni (1969: 142) suggests, this formation should be interpreted in an opposite way. The forest thicket was established by farmers in what were earlier baobab-rich savannas. Savannas were upgraded to form thicket using special weeding practices, such as weeding before the burn so as to limit the destruction caused by bush fire. According to Mondjannagni, the baobab savannas, which have virtually disappeared, have thus transformed into baobab wooded thicket. The thicket is therefore anthropogenic and contains young vigorous forest trees such as *Mallotus oppositifolius*, *Antiaris africana*, *Dracaena arborea*, *Dialium guineense*, *Uvaria chama*, as well as savanna trees such as *Fagara xanthoxyloides* and *Dichrostachys glomerata*.

Brouwers (1993) gives an insight into the farming skills and knowledge in the region where this vegetation had been common. He details how farmers are today enriching their soils through the use of manures, crop rotation and oil palm fallowing. For example, farmers here find they need to 'wake up' soils which are or have become unproductive. They do this either by planting soil-enriching crops, by adding organic and now even inorganic fertiliser, or by encouraging palms to take over the field as a fallow crop. This is associated with an enhancement of worm and termite activity, the latter possibly being the subject of management (see Iroko 1982). Yet while Brouwers is highly sensitive to such local management issues, he is nevertheless entrapped within a broader analysis of long-term forest degradation. While he documents the demise of bush fallow, he appears to have ignored its earlier creation from grassland. Like Aubréville, he suggests that the 'original natural vegetation' was sub-Sudanese clear forest, with tree species like *Milicia excelsa*, *Parkia biglobosa*, baobab and others.

The hypothesis can be forwarded that when, in the nineteenth century, land pressure in this region was low, and newcomers could still gain access to land, farming would upgrade the savannas into these baobab thickets used for bush fallowing, thus creating the thicket which Aubréville and Mondjannagni noted. But many areas of bush fallow declined in the mid-twentieth century, and by the time of Brouwers' research in the late 1980s bush fallow had disappeared. With population densities now between 80 and 250 persons per km^2, farmers have developed more intensive fertility management regimes including short cassava and pigeon pea fallows, and especially oil palm-fallow, mixed with shrubs.[3]

The presence of baobabs in the sacred forest of Foncomé should have alerted Aubréville to one of two possibilities. One was that this sacred forest was, in fact, an old village site; in such sites baobabs are frequently established. The forest patch was thus growing over very particular soils (the presence of baobabs has been observed at the heart of anthropogenic forest islands in Guinea (Fairhead and Leach 1996a). The second possibility is that baobab savannas might have been characteristic of the region; and that savannas might have ceded to a forest vegetation.

As we noted early in this chapter, all analysts of Benin's vegetation stress unrelenting forest decline. The FAO (1981) forest assessment for Benin accepted without critical reflection Aubréville's assertion that he was observing in the 1930s the felling of the last stands of forest in the south and of forest islands in the central zone. Yet as we have seen, Aubréville's reading of the landscape may have been erroneous. Not only was it inconsistent with Chevalier's observations, but also none of the indicators which he considered to represent recent forest loss can be accepted without enquiry. Significantly, FAO (1981) suggests that these 'last vestiges of forest' which Aubréville and Chevalier saw, and which they thought were disappearing in 1909 and 1937 respectively, are still disappearing. The question arises – how long can 'continuously disappearing last vestiges of forest' last until they do finally disappear? Have last vestiges been continually disappearing for a century? Puzzlement increases when we note that in a 1924 forest assessment, Mangin put a figure on Benin's moist forest of around 50,000 ha (Mangin 1924). After some sixty years of supposed incessant deforestation, FAO suggest that almost the same area – 47,000 ha – of forest existed in 1980 (FAO 1981).[4]

Vegetation cover in earlier centuries

We have argued that forest islands and sacred groves cannot necessarily be considered as forest relics. Equally, tall forest trees in their forest form can grow in open fields, and need not, therefore, indicate the recent presence of denser forest. We suggest also that one cannot assume that thick forest thicket as found in fallows or indeed palm groves is necessarily degraded forest. If these forms of vegetation were all that was 'left' in Benin in 1900, then the question of Benin's earlier forest cover becomes more open.

The nature and extent of Benin's deforestation in pre-colonial times have been the subject of recent heated debate. Gayibor (1986, 1988) has affirmed the conventional view also held by Chevalier and Aubréville (in his early years) that people had savannised the coastal forest between the sixteenth and twentieth centuries. This position has, however, drawn strong criticism from Blanc-Pamard and Peltre (1987). While Gayibor used the accounts of early European travellers, supplemented by some oral evidence, to suggest the gradual deforestation of the zone, Blanc-Pamard and Peltre interpret the same data as indicative of a continuity in the mosaic of vegetation forms.

One of Gayibor's earliest sources is that of British navigators of 1588: 'Once Elmina fort is passed, the coast presents only large forests, sometimes so thick that it would be impossible to penetrate.' He suggests that Prevost (1746: 1, 21–2), Dapper, Bosman, Labat, Isert and others (all key sources for pre-colonial West Africa) all describe a richness of soil and vegetation. Yet as Blanc-Pamard and Peltre point out, most of the authors that he cites give descriptions only of farmland and palm groves, not of forest (e.g. Isert 1788 in the Aneho hinterland and Watchi country; Robertson 1819 in Ewe country). On the coast, in citations of Dapper, no mention is made of woodland or forest, but again only of productive open farmland: 'a land of plains and beautiful valleys, cut by large rivers' (Dapper 1668: 303–4). The citation of Bosman (1705) is similar, describing a farmed land of 'verdure', a singularly ambiguous term often used to describe grassland. The citation of des Marchais (in Labat 1730) also makes no mention of forest. Indeed, as Blanc-Pamard and Peltre point out, in these texts forest is often notable for its absence. Smith – who is the only author cited to suggest that there was a 'prodigious quantity and infinite variety of beautiful high trees'– also qualifies this by noting that 'they seem to be planted express to serve as adornment' (1744, II: 133–6). Smith, one should note as well, goes on to describe equally eulogistically the adorned landscape as completely farmed and populous. Further citations from Isert in no way indicate the presence of dense forest, and where they describe trees suggest that these were planted as in the citation given earlier.

Apart from these works, Gayibor cites only secondary and unpublished oral historical accounts by Gaba (1942: 141), who suggests that Aneho and the interior were forested when Ga migrants arrived. There are, in fact, more oral accounts which Gayibor could have used in support of his argument (e.g. Pazzi 1979: 161). Nevertheless, the presence of 'forest' in accounts of oral history need not suggest the ancient presence of dense moist forest in the region. In Benin, as elsewhere in the transition zone, derelict palm groves are often described as 'forest'. Unless one knows what is meant by 'forest', these oral accounts are extremely hard to interpret. As Blanc-Pamard and Paltre (1987: 421) suggest, in relation to the oral evidence produced by Gayibor, the account is too imprecise in time and space and anyway contains within it problematic internal contradictions.

Gayibor also uses certain place names to suggest that villages were founded in forest (*Ave* – 'forest'; *Avenu* – 'at the boundary of the forest'; *Avepozo* – 'burnt forest'; *Avedji* – 'on the forest'; *Zu* – 'forest': names in the Aja language). He argues that as all these localities are now in savanna, they indicate deforestation. Wartena (1992) also cites certain other names in the Lama depression as evidence for the presence of forest when villages were founded.[5] Yet Blanc-Pamard and Peltre reasonably ask in their rebuff of Gayibor whether the place names indicate more the persistence of a vegetation mosaic than the early presence of continuous forest: 'Why would villages situated in dense forest carry names explicitly referring to this relatively homogeneous milieu? . . . in an area more contrasting of a mosaic of forest and savanna, one would understand better why some of them distinguish themselves by reference to their situation in or in proximity to a forest island' (1987: 421). In 1988 Gayibor had the opportunity to respond to all of these criticisms, but refuted none of them. The debate must remain open, but it revealed that there was as yet no compelling evidence for the presence of continuous stretches of high forest cover along the coast of Benin.

To examine early vegetation, we thought it instructive to compare the more detailed village-to-village accounts of those travelling from the coast at Ouidah to Abomey, the Kingdom of Dahomey. While there is little documentation on travel to the interior pre-1890, descriptions along this particular trade route provide an exception, as it was relatively well trodden by European travellers from the 1770s onwards. There are, of course, limits to the generalisability of this endeavour as it was a major trade and political route, with its vegetation subject not only to the demographic implications of trade, but also to the explicit vegetation management of a military route. The journey passed through the forest of Lama. Until Heve, the route is indicated on the 1893 map. The landscape descriptions made on this route, summarised in Table 5.3, are hard to interpret. Nevertheless, 'woodland'

Table 5.3 Landscape descriptions made on the Ouidah to Abomey journey

Location and 1893 map description	*Norris 1789*	*Duncan 1847*	*Freeman 1844*	*Burton 1864*	*Comparison*
Ouidah					
Open with groves of trees.	Open, interspersed with groves of lofty and luxuriant trees.	Open	Open country	Open; tall savanna; clumps of palm trees and scattered palms; copse near water courses.	No change
Salwi/Savé					
Open; tree groves.		Level; wooded for the last stage.	Large savannas; small clumps of trees.	Bush land; a few clearings, so thick axes needed to leave the path. From Denun: grass, bush, thin palm forest, no shade. Tall grass, wild trees.	More wooded
Torri					
In wood extending 2 km sw.	Nr river, banks with stately trees. Much underwood shelter to plenty of elephants.				
Densely wooded no grass.	Thick wood; grass higher than heads; heat fatiguing.	Last stage wooded path.	Less open; low brushwood.	Grass disappeared except in the farms clearing.	Less grassland?

Asowa(y)					
Dense vegetation; some clearings.		As park in England.	Forest island.	Forest island. Lane of shrubbery; sundry long flats; well-wooded ascents.	Grass become bush fallow?
Alada					
Dense vegetation.	Many palm trees nearby.	Thickly wooded; immense large and straight trees.		Alternately grassy and bushy, had been burnt in the dry season.	
Havée					
Dense vegetation.		Thickly wooded; high trees.		Surrounded by giant trees. Impenetrable herbaceous growth; noble forest trees (e.g. 80 ft cotton trees).	
Whybo					
		Beautifully wooded.		Trees more gigantic.	
Appa Akpway					
	Great wood commences.	Swamp; trees of immense size interspersed with smaller fruit trees.	Marsh 8–10 miles; low brushwood; from Avadi, grass, burnt in places.		
Agremi					
	Beyond wood.	Open, tree clumps resembling park.			

Table 5.3 Continued

Location and 1893 map description	*Norris 1789*	*Duncan 1847*	*Freeman 1844*	*Burton 1864*	*Comparison*
Kwo		Open to afar.	Open, travel, under sun. Savannas and odd forest tree.	Open with gallery forest.	
Kanna	No trees.		Baobab, *Inga lugubrosa*, Guinea peach, *Sarcocephalus esculentus*, palm tree.		

usually refers to tree cover in savanna and hence not to forest. A vegetation of high trees in woodland is thus not to be confused with what we are considering as forest. Forest, such as that existing around villages, was usually easily distinguishable from it. 'Groves of trees' often refers to palm tree groves, but might also be fetish groves or forest islands.

From the differently dated descriptions – whose authors had no axe to grind concerning vegetation cover, so one has no reason to suspect overt biases – it could be argued, first, that the land from Savé to Torri, and from Torri to Assoway, and Assoway to Allada became more wooded over the period 1772–1890, but that from Allada to Whybo there is ambiguity. Despite the terms 'thickly wooded' and 'immense trees' being used, this may not be excluding grassland, but grassland does seem to have been excluded on the 1893 map. The marshland forest of Lama between Appa and Agremi, crossed by Freeman in the 1840s, might have suffered some event – possibly a forest fire – between then and Duncan's journey only a few years later. Their two descriptions appear irreconcilable.

Burton on his journey noted a derelict landscape:

> [T]he aspect of the country confirms the general impression that the Dahomans were . . . an industrious race, till demoralised by slave hunts and by long predatory wars. The land has recently been well cleared, and it is still easy to reclaim, though in time the fallows will be again afforested. Others opine that it has of late been royal policy to gird the capital [Abomey] with a desert, as the surest defence against invaders . . . a ruined region . . . beyond a few towns, the country is a luxuriant wilderness.
>
> (Burton 1864: 116)

This image of depopulation and dereliction is supported by many historians of the region and here we will not enter the debates in detail. Three points would certainly be central to a more detailed examination of pre-colonial Benin. First, the coastal region had immense populations around the 1700s. When the coastal town of Ouidah was the major political force in the region, the area appears to have been heavily populated.

> The country is so populous and full with so many villages that the entire state [of Juda] appears as a very large town, divided into many quarters, separated from one another by land cultivated with care, which seem to be only gardens, of which the soil is of an incredible fertility. There is not the least bit of land derelict or neglected, and that without discontinuing, and without giving the land the least rest. The inhabitants look after their land so carefully that the paths are everywhere are tiny streets.
>
> (Labat 1730, cited in Études Dahoméenes xvi: 76)

Just how far inland such population densities extended would need to be considered. Second, this port was exporting some 18,000 slaves per year from the 1700s and slave exports continued well into the nineteenth century. The origin of these slaves and the political, military and demographic fall-out would need to be taken into account. It becomes clear why forest islands were important as fortresses. Third, the major shift in power from Ouidah to Abomey, its demographic impact and the population policy of the ascendant Abomey regime would need to be explored. While more detailed historical research would be needed to explore these issues, it is certain that vegetation change in this region cannot merely be conceptualised as the 'gradual savannisation of forest due to rising population and increased farming'. Rather each area's landscape will have a vegetation history linked closely to the region's more turbulent real history.

Conclusion

While limited in the precision with which it can detail vegetation change, this chapter has attempted to raise several questions concerning the way in which vegetation change in Benin has been considered. The history of vegetation in Benin appears to be something other than a history of relentless deforestation from an 'original' forest, whether during this century or over the last few centuries. We have, we think, falsified assertions concerning massive forest cover early this century and have encouraged those who put this process back a century or two further to reconsider their interpretations of their data.

Within Benin, it must be considered that all three forms of anthropogenically enriched landscapes – forest island, palm groves and baobab thicket – might have been read backwards by specialists, who consider each of them either as degraded forms or, in the case of forest islands, as relic islands of conservation which suggest an otherwise decimated landscape. If, as has been shown to be the case elsewhere, the sacred forests of Dahomey are found (a) to have been planted, or (b) to have become established on old village sites, or both – that is, they are anthropogenic, or on anthropogenic soils – then it would seem that the use of these forests as testimony of a now lost natural vegetation would be incorrect. If, as is the case, dense palm groves are established in savanna and if, as is the case, baobab bush is not degraded forest but enriched savanna, then the whole edifice of suppositions concerning 'Benin's past vegetation' begins to look very weak. In fact, a totally different hypothesis for vegetation change could be forwarded.

At the time when Chevalier and Aubréville laid the grounds for analysis of Benin's vegetation, they considered that the West African climate was in the process of gradual desiccation. Aubréville actually argued that the climate of Benin was too dry at present to support equatorial forest all along the coast of the Dahomey Gap which would enjoin the major forest blocks on either side. Nevertheless, he argued (following Chevalier) that only a few centuries ago, the Upper Guinean forest west of the Dahomey Gap and the equatorial forest to its east were linked by a thin band of coastal forest. He suggested the climate had earlier been around 20 per cent more humid because of the supposed beneficial effects of the 'original' forest cover on rainfall. Thus the forest across the Dahomey Gap, he argued, had maintained the conditions for its own survival. Following forest loss, the climate had subsequently become as dry as today's. Today's climate was no judge of past forest extent. It was, in part, in the context of climatic desiccation that his interpretations of isolated trees, forest islands and baobab thicket in terms of degradation made sense. Crucially, however, Aubréville later came to see that West African vegetation was responding to long-term rehumidification, rather than desiccation. Yet neither Aubréville nor his intellectual descendants have reconsidered their precepts concerning Benin's vegetation in this light. In this chapter we have attempted the beginning of such a reconsideration.

Notes

1 Isert also went on a one-day elephant hunt in the grasslands around Ouidah. 'After a long walk through the dewy wet grass, on three different occasions I had the pleasure of seeing small herds of these majestic animals.' When shot, 'they ran into the thickest bushes, where no one could follow them. Small trees which were in their way when they were running away were uprooted without much ado. The elephants paid no attention to the many thickets which would have torn the hide of other animals, but moved as lightly and proudly through them as if they were on a carpet' (Isert 1788/1992: 104). He notes that the elephants were only seven or eight feet tall.

2 It should be noted that FAO authors also claim to have identified 4,450 ha of degraded semi-deciduous forest in 1949 in the Djougou region. Yet this could not be ground-truthed. Furthermore, Aubréville in 1937 was explicit in suggesting that the forest islands owed their existence to site-specific soil-water factors and were wholly different from the drier forests in their vicinity (pp. 24–7). One must assume that what FAO/UNEP/GRID 1980 took to be degraded semi-deciduous forest was something else. Only 62 ha of degraded forest remained in 1975, and the authors are unclear whether this might actually be degraded forest island.

3 This, Brouwers argues, is as effective in restoring fertility as bush fallow, if not more so (see Brouwers 1993). Indeed, despite huge population increase and capital extraction, innovations such as these have kept agriculture relatively stable in southern Benin over the last century (Pfeiffer 1988; Brouwers 1993).

4 More recently, Sayer *et al.* (1992) have estimated the forest area at 35,500 ha. While supposedly based on satellite imagery, this figure cannot be considered as any more accurate than those above, as the coarse-resolution NOAA/AVHRR

satellite data used were not capable of resolving forest areas; while other satellite imagery was used, it was not ground-truthed in Benin.

5 We might add that Wartena questions whether any of the plateaux of Allada, Adji and Aplahoué carried forest in historical times, as they lack village names bearing the term 'forest' (Wartena 1992).

6

TOGO

Introduction

According to the World Conservation Monitoring Centre (Sayer *et al.* 1992) the original forest cover of Togo was 1.8 m ha, compared with a present cover of 136,000 ha. But while virtually all analysts of Togo consider forest to be in steady decline, none considers the bulk of deforestation to have taken place during the twentieth century. In this chapter we examine issues surrounding forest cover change since around 1900 and then go on to examine how pre-colonial forest cover change has been analysed – in the case of Togo, in relation to pre-colonial iron smelting. The emphasis here, then, is less on a critique of modern forest statistics than on the ways that the past existence and decline of forest has been mutually framed by historical perspectives on economic and technical change. The extent to which Togo's pre-colonial forest history is one of an emerging 'woodfuel gap' in the context of ironworking by growing populations invites reappraisal in the light of the challenges to this perspective recently forwarded for modern times (G. Leach and Mearns 1988; Dewees 1989). (See Map 6.1.)

Forest cover change in the twentieth century

In 1911, when the British forester Unwin visited Togo (then under German control), he concluded that the country had less than 1 per cent forest cover.[1] The German forester Metzger (1911) wrote that only 1.5 per cent of the original forest remained at the turn of the century, the rest, he argued, having already been converted to savanna.

Zon and Sparhawk (1923) used Metzger (1911) to calculate that Togo's forest (dense humid forest and riverside forest) in 1923 covered 138,000 ha (or 1–2 per cent of the country), of which only 60,704 ha were not riverside forest. As c. 60,000 ha of this forest were in the region which later became part of Ghana, the corrected figure for today's Togo borders would be in the region of 78,000 ha. This figure is consistent with

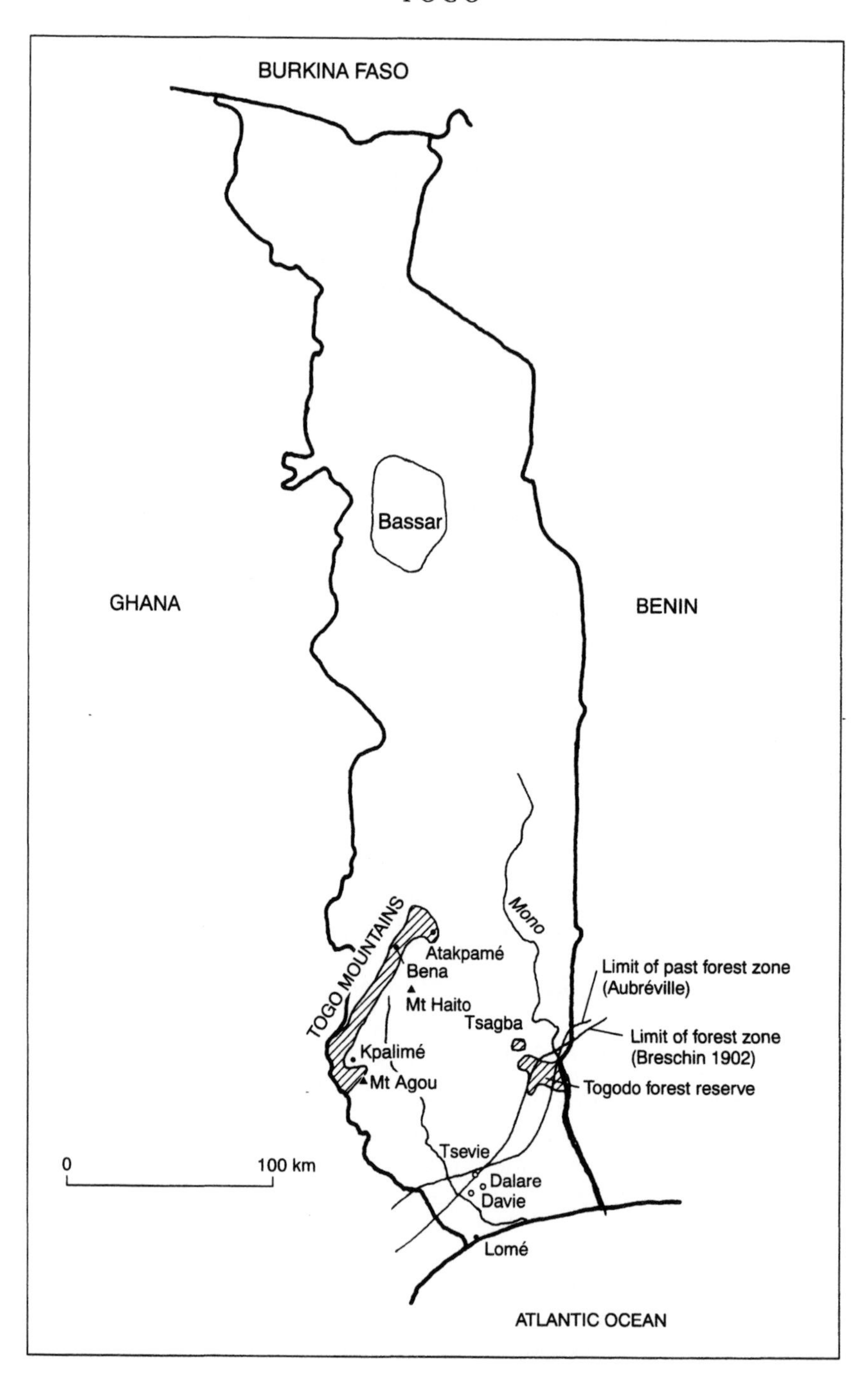

Map 6.1 Regions of Togo described in the text.

Aubréville, who examined the forests of what had become French Togo, suggesting that 'the equatorial forest represents only a few tens of thousands of hectares, a tiny proportion of the country' (Aubréville 1937: 35).

These estimates of forest cover in Togo are also supported by the travel accounts and maps left by early German visitors to Togo. Early observers of the coast and immediate hinterland did not describe forest cover. Thus in the descriptions of Zoller (1885) of the coastal strip, the only 'high' forest was to be found along streamsides, the rest of the landscape being either fields or baobab savannas.[2] The British forester Unwin described this landscape as consisting of thorny shrubs and occasional baobabs (1920). Early observers did identify one area of 'forest cover'. The Basel missionary, Bürgi, toured Ewe territory, and found actual 'forest' only round Davie and Darave [Dalave]. This, however, was palm forest, 'where the oil palms and silk-cotton trees are so thick that one can walk for hours in shadow' (Bürgi 1888). Yet given the predominance of palm trees and silk-cotton, both of which were commonly established in savanna by farmers investing in what was then the highly profitable palm oil economy, it would appear that this 'forest' could have been anthropogenic and established in savanna. This would be analogous to the anthropogenic palm forests known to have established in savannas during the mid-nineteenth century near the Volta river, just across the border in Ghana (see Johnson 1964: 21), and in Benin, as we have described in earlier chapters.

Taken at face value, the more recent estimations of Togo's forest cover do not show a decline since estimates made early in the twentieth century; they even suggest the contrary. Aubréville (1956) estimated the area of the wet forest zone to be 200,000 ha and the area of forest to be 180,000 ha. Persson (1974) estimated Togo to have 380,000 ha of closed equatorial forest. FAO (1981) estimated that Togo had perhaps 304,000 ha of forest, a figure which included forest cover over plantations, so that 98,000 ha were considered to be undisturbed. Sayer *et al.* (1992) put the figure for lowland rainforest at 136,000 ha.

While all these figures are higher than the forest cover estimates made early in the century, the authors of the modern figures suggest that Togo's forests are in decline. FAO (1981), indeed, suggested that deforestation was occurring at around 2,000 ha per year (1976–86), a figure accepted by Sayer *et al.* (1992). The FAO 1990 Forest Assessment (FAO 1993), which incorporates a very different definition of forest, nevertheless reinforces this idea of forest loss. With 1.3 m ha of Togo thought to have been forest in 1980, 22,000 ha are considered to have been lost annually, at a rate of 1.6 per cent/year.

This inconsistency needs to be explored. Figures for forest cover appear to have increased at a time when, it is argued, the forest is in decline. It could be argued that the problem lies in discrepancies in defining what is forest. Yet the work of Unwin and Aubréville suggests that forest as they defined it, and forest as FAO (1981) defines it, were more or less the same. A second possibility is that modern estimates are simply more accurate, given the measuring procedures now available using air photographic and remote sensing possibilities. For many, these two explanations are adequate. Furthermore, those forwarding them can point to evidence of deforestation. Some of this evidence is somewhat spurious, however. For example, FAO (1981) suggests that forest tree species growing in their forest form in fields, as observed in much of Togo, are evidence of forest loss – a deduction which we have found much cause to criticise in other chapters. Others note areas where forest cover has incontrovertibly declined and suggest that this process is generalised; by deduction they argue that early figures seriously underestimated forest cover.

The best cases demonstrating forest decline are those examined in FAO/UNEP/GRID (1980). The authors of this study compared air photographs and satellite images of assorted dates to estimate forest cover change in two sites between 1949 and 1969 (FAO/UNEP/GRID 1980). Furthermore, the authors estimate that 80 per cent of forest area in the site of Togodo (eastern Togo) in 1969 had been lost by 1977 (see Table 6.1).

These are clearly examples of deforestation, but they are certainly not representative. Indeed, it could be argued that they are highly exceptional. The researchers sought to compare sites where there had been significant forest loss, but they found this much harder than expected. As they pointed out:

> From preliminary discussions with representatives of the government agencies concerned and with technical assistance personnel working in the countries, it seemed as if the main task would be to make a [site] selection among suitable study areas [where there had been major loss], rather than their identification. In practice, however, the main problem turned out to be the location of areas where significant changes had occurred and were displayed on available air photography.
>
> (FAO/UNEP/GRID 1980: 36)

Table 6.1 Forest loss in two sites in Togo 1949–77 according to FAO/UNEP/GRID (1980)

Site	*Forest area*			
	1949	*1969*	*1977*	*ha lost/year (1949–69)*
Togodo	9,537	8,400	c. 1,400	57
Tsagba	6,237 (+681 degraded)	238 (+ 406 degraded)		300 (+14 degraded)

Given the small scale of the area of Togo's forests, it is possible to examine forest cover change issues on a case-by-case level. The forests examined above are the 'last remnants' of what are classified as coastal semi-deciduous dry forests (FAO/UNEP/GRID 1980). Furthermore, it would appear that they are growing over areas which were, in earlier centuries, populated (Cornevin 1969). Moreover, it may be that these forests have only relatively recently become closed formations, containing within them as they do mature specimens of savanna species such as *Ficus* spp., *Sterculia tracagantha*, *Phyllanthus discoideus*, *Daniellia oliveri*, *Butyrospermum paradoxum*, *Bridelia* sp., *Antiaris africana*, *Albizia zygia* and *Elaeis guineensis* (FAO 1981: 522).

Apart from these forests, semi-deciduous and deciduous forest formations are found in only three other locations: (a) as peri-village forest islands and gallery forests, in the upper basin of the Mono and Ogou rivers; (b) on the banks of the Mono river; and (c) on the slopes of the Danyi plateau, and mount Togo, Haito and Agou mountains. In these areas, the forest is highly dispersed, with hardly any patches larger than 5,000 ha.

None of the peri-village forest islands and gallery forest in the upper basin of the Mono and Ogou rivers are more than 500–600 ha (FAO 1981). These forest patches are similar to the forest islands across the border discussed in Chapter 5 on Benin. As elsewhere, one can ask whether these forest islands are 'natural', or whether they were established around villages inhabited, vacated or otherwise, or established on abandoned village sites. While several of these forest islands occur in the almost uninhabited reserved forest of Abdoulaye, this region had been inhabited in the past (see Cornevin 1969, first edn: 280–82). Indeed, it may be that the forests in this zone indicate where old settlement sites were – a finding which would be in keeping with analogous conditions around the depopulated region of old Oyo in Nigeria (see Keay 1947; Abimbola 1964).

Semi-deciduous and deciduous forest formations also lie on the mountain ranges from Atakpamé to Kpalimé. In 1937, during his forest assessment and policy formulation mission to Togo, Aubréville examined these equatorial forests and did not find them intact. He asserts that such forest as remained was maintained 'only on the slopes, in the valleys and on the foothills on the eastern side, and on the abrupt flanks of the mountain' (1937: 34). The convex parts of the slopes were denuded, or almost, whereas the concave parts remain wooded. The road from Atakpamé to Kpalimé, which follows the base of the range, held more or less to the border of this forest. The upper forest border on this eastern slope attained the watershed and summits and sometimes redescended on the opposite flancs, but often stopped on the ridge itself. The peneplain extending south-east, in front of the Togo mountains, consisted only of poorly wooded savanna 'as far as the eye could see' (1937: 34).

Aubréville suggested that the forest once (but no longer in 1937) formed a continuous band from Atakpamé to Kpalimé, and that perhaps long ago the forest even covered the plateau. He noted that the plateau was (in 1937) covered with woody savanna and, towards the south, by open tall-grass savannas. The only forest areas on the plateau were the gallery forests which criss-crossed it. He presumed that the tall-grass savannas indicated 'the ancient existence of the forest'. In general he asserted that the forest was being heavily attacked by fire and indigenous cultivation and was in regression, but he had no historical data to prove this.

In Aubréville's opinion certain of the small forests were 'very beautiful', and he described those in: the valleys of (a) Amoutchou, which go up to the Atakpamé Okou road; (b) Houlé, which starts near Esimé; and (c) Amouoblo, occupied by 'very beautiful stands', which still exist in the upper basins, on the ridges and in the valleys, despite being 'heavily attacked'. South of the Togo Mountains, near Kpalimé, in the ravines and on the slopes the equatorial forest also existed 'in all its splendour', in the wooded area of Misahoho and the ravines of the Egoui and of Etchi which follow the route from Klouto to the frontier of the Gold Coast. Aubréville urged their reservation.

Aubréville described these forests of the zone between Atakpamé and Kpalimé as in the process of degradation. This same forest is present, although considered always to be 'in the course of degradation' (see plate 6.1 in Guelly *et al.* 1993: 37). FAO (1981) among others considers that the cocoa and coffee grown in much of this forest clear the undergrowth and prevent regeneration.

Nevertheless, in a simple but superb article – the most recent research to come out of these mountains – a radically different set of possibilities is suggested. Guelly *et al.* (1993) argue in the conventional sense that farmers degrade and savannise the forest and are then forced to farm other lands which are already savannas. Yet the article goes on to show how farmers can and often do install a forest formation in these savannas – creating anthropogenic forest fallows. 'The cultivated savannas, when abandoned, evolve naturally into forest' (Guelly *et al.* 1993). They note the succession from fields to pioneer trees of *Harungana madagascariensis* (also *Trema orientalis* and *Vernonia colorata*); thence after 10 years *Margaritaria discoidea* (among other species) comes to dominate; after 20 years *Macaranga barteri*; and after 25 years the forest becomes mature with *Pycnanthus angolensis*, *Canarium schweinfurthii*, *Milicia excelsa* and *Erythrophleum* sp.

Although Guelly *et al.* assume that the forest which was originally farmed was 'original forest', they do not prove this. The validity of their assumption is clearly undermined, as the authors do show that farmers are creating forest out of savanna. This finding throws the history of vegetation change in this region open to much wider reinterpretation. Could it be that much of the forest land on the mountains and plateau which they argue has recently been savannised was itself originally savanna land which had been converted to forest?

It is noteworthy that Guelly *et al.*'s (1993) study of forest vegetation took place in the vicinity of the village of Bena, situated on the plateau which Aubréville in 1937 described as savanna. Careful analysis of vegetation change in this region of Togo using old and new air photography (or remote sensing data) might provide a key study in assessing and comprehending West African vegetation change.

The results of Guelly *et al.* are more radical than the authors suggest. It is generally held that the climate in Togo is marginal for forest cover. In this context people are considered only to degrade forests: as FAO (1981) put it, the marginal climate makes these forests 'very vulnerable to anthropogenic disturbance'. But what if disturbance can create forest, as Guelly *et al.* show, and as we have found elsewhere in the transition zone (Fairhead and Leach 1996a; see Spichiger and Blanc-Pamard 1973)? Are the savannas in this region not equally vulnerable to anthropic disturbance which constructs forest? That the forests of the Togo mountains have a very limited species diversity (FAO 1981) might suggest that they are more recent than has been supposed.

Given (a) that it has been shown that people can establish forest fallows in savanna, (b) the importance of forest cover for tree crops and (c) the importance of tree crops such as cocoa and coffee in this region during the periods when prices have been good, it would appear reasonable to suspect that farmers have upgraded savanna lands to create forests to harbour tree crops. If this is the case – and only detailed time-series research would be able to show whether this is so – then it could be imagined that far from diminishing, forest cover in these mountains as defined by FAO 1981 (in which half of the 304,000 ha of forest harbours tree crops), or as picked up on satellite imagery as forest by Sayer *et al.* (1992), may indeed have increased. Such conclusions may sound far-fetched but, as our analyses of Côte d'Ivoire, Ghana and Benin have shown, would be in keeping with findings elsewhere in the transition zone.

Pre-colonial vegetation change

Without exception, as we have noted, those who have expressed opinions concerning vegetation history in Togo over the last few centuries have spoken of forest decline. Aubréville (1937) suggested that southern coastal Togo had once been covered by a dense xerophile (dry) deciduous forest, but was now a landscape of oil palms, crops and secondary bush. Metzger (1911) argued that this coastal region had once been analogous to the closed-canopy forest of Cameroon. His pre-colonial history of decline is accepted by more modern authors, such as Gornitz and NASA (1985). But most notably in Togo it has been supported in the work of Goucher (1981, 1988), who has highlighted the supposed role of iron smelting in degrading the forests. Goucher's work is somewhat compelling and has been influential in studies of West African forest history (e.g. Nyerges 1988). Indeed, she has made generalisations concerning deforestation throughout the West African forest belt from arguments based partly on her analysis of Togo (and Ghana). It is therefore important to examine Goucher's arguments in some detail.

Goucher argues that the evidence from archaeology, oral history and early travellers' written accounts confirms that there was extensive loss of forest prior to the twentieth century. She attributes this partly to climatic desiccation during the sixteenth to eighteenth centuries, partly to farming, and partly to the fuel demands for iron smelting. The earlier presence of more abundant forests is merely assumed, although she also illustrates the assumption with two lines of argument. First, she argues that large forest areas existed north of the ninth parallel on the basis of place names. Names such as *Lama* and *Lanmba* mean 'those of the forest'; *Lan*, 'forest', and *Lawda*, 'in the forest' to indicate that the region was once characterised by forests. Apart from asking 'what is meant by the local language term for forest?', the further question that Blanc-Pamard and Peltre (1987) asked of Gayibor's use of toponyms, which we outlined in relation to Benin (Chapter 5), remains significant: why name villages after forests, if all around is forest? Such a name is more noteworthy if much of the landscape is not forest; indicating the presence of a patchy vegetation, probably of forest islands in savanna. And as mentioned above, this region was[3] and indeed remains dotted with small, supposedly 'relic' forest islands. As these were each associated with a village, we argued, they were very probably anthropogenic forests, less the relics of a forest once destroyed than the product of inhabitation either by present inhabitants and their ancestors, or by earlier populations which had subsequently vacated them. Second, apart from place names, Goucher also infers that German colonial alarm about deforestation was evidence for it. This is a bold assumption. As we have argued in Chapter 1, administrative alarm at deforestation generally preceded any analysis of it.

Apart from these two sources, Goucher provides no further evidence for the past existence of forest. Yet she isolates two major causes of deforestation. In accordance with orthodox views, she argues that agricultural intensification 'led to the widespread conversion of woodland to grass savanna'. While this captures received wisdom, she adds no further data and, as we have argued here, this position is not without counter-evidence. It overlooks how many farming practices in the transition zone increase, not decrease, woody cover (Blanc-Pamard 1978, 1979; Mitja 1990; Guelly *et al.* 1993; Amanor 1994b; Fairhead and Leach 1996a). Such assertions need to be contextualised by asking: what sort of farming, in what period, and in relation to what prior vegetation state?

But above all – and now we get to the crux of the argument – Goucher (1988) contends that the fuel requirements for iron smelting were even more destructive than agricultural practices. Iron smelting dates back some

2,000 years in West and Central African history. She suggests that parts of Togo were the hub of a large complex of iron and steel smelting, when even during its decline in the late nineteenth century, some '500 furnaces were seen operating at a single smelting site'.

Her conclusions for Togo draw heavily on an earlier work concerning the deforestation effect of iron smelting more generally in West Africa. First we shall examine Goucher's general argument that charcoal making for iron smelting was largely responsible for deforestation. Then we shall examine more closely its application to Togo, as the argument has been adopted by other analysts there. Her arguments strongly echo analyses of the 'woodfuel crisis' in modern times, which hold that under increasing woodfuel demand, a gap tends to emerge between fuel consumption and the regenerative capacity of trees for supply.

Iron smelting in West Africa

Goucher (1981) argued (a) that smelting was a major cause of deforestation;[4] and (b) that among other reasons West African smelting declined as a result of wood scarcity (rather than foreign competition, as is often supposed). She argues that only a few savanna tree species produce charcoal of good smelting value.[5] Without the proper charcoal, iron could not be made, and proper charcoal was made only from slow-growing species. The species she lists for Togo (so specific to the industry as to be quasi-secret) are all common savanna species: *Combretum* spp. (a savanna shrub), *Detarium senegalense* (a savanna tree, of which one variety grows large in forest outliers) and *Crossopteryx africana* (a savanna tree). She notes in particular the use of *Burkea africana*, *Acacia* spp., *Prosopis africana* and *Zizyphus mucronata* (Goucher 1981).

The preferred species are common to the savannas in the region, but they represent only 7 out of some 60 of the species found there (see Aubréville 1937). Goucher argues that the scale of iron smelting required a 'staggering' number of trees to supply the charcoal, using the Ghanaian site of Dapaa where slag volumes have been assessed to support her case. Yet if one calculates area of land required to supply this fuel, we think it can be argued that the fuel can be sustainably produced within the landscape. To make a rough estimation of the area of land required to produce the fuel, we must assume – incorrectly as we have seen – that all savanna trees are suitable for smelting, even if some are more preferred than others. Were we not to assume this, then it would be hard to argue that smelting engenders deforestation, as it would lead only to the depletion of these species, not of woodland in general – a wholly different issue which Goucher glosses over.[6]

Goucher indicates that the site of Dapaa (in the forest–savanna transition zone in the centre-west of Ghana) contained about 210 tonnes of slag which, she states reasonably, would require c. 840 tonnes of charcoal to produce. Using a low 10 per cent conversion rate from wood to charcoal, this implies the use of 8,400 tonnes of wood. She calculates that if an average *Burkea africana* tree weighs only 27 kg (i.e. not a mature individual, as a mature *Burkea* can be up to 15 m high and 1.8 m in girth, weighing perhaps ten times as much[7]), a little over 300,000 trees would be needed over the period when smelting was in operation on site.[8]

But Goucher leaves the calculation at that, without estimating the area required to produce this 'staggering' number of trees. From Table 6.2 it is possible to see just how small such an area is, even in relation to the area of a small village territory (say of 2,500 ha, a circle of less than 3 km round a village), even if it were only partially wooded. As Table 6.2 also shows, the area needed becomes much smaller if – as was the case – the smelting took place over a longer period, therefore making use of subsequent tree regrowth.

In interpreting the table, one might note that Goucher cites Hopkins (1965) to suggest that 'the number of individual trees per acre of "derived

Table 6.2 Range of areas required to produce 300,000 trees under different densities over different timescales

Original desnisty of trees (average spacing)	*Number of tree growth periods during which smelting operated (assuming savanna tree regrowth to be over 30 years)*				
	1 *i.e. all trees felled within 30 years*	*2* *i.e. trees felled over 60 years, returning once to regrowth*	*3* *i.e. trees felled over 90 years, returning twice to regrowth*	*4* *i.e. trees felled over 120 years, returning three times to regrowth*	*10* *i.e. trees felled over 300 years, returning ten times to regrowth*
2 m × 2 m (2,500/ha)	120 ha	60 ha	40 ha	30 ha	12 ha
3 m × 3 m (1,111/ha)	270 ha	135 ha	90 ha	68 ha	27 ha
4 m × 4 m (625/ha)	480 ha	240 ha	160 ha	120 ha	48 ha
5 m × 5 m (400/ha)	750 ha	375 ha	250 ha	187 ha	75 ha
6 m × 6 m (277/ha)	1,083 ha	541 ha	361 ha	270 ha	108 ha
7 m × 7 m (204/ha)	1,470 ha	735 ha	490 ha	367 ha	147 ha
8 m × 8 m (156/ha)	1,923 ha	961 ha	641 ha	480 ha	192 ha
9 m × 9 m (123/ha)	2,439 ha	1,219 ha	813 ha	609 ha	243 ha
10 m × 10 m (100/ha)	3,000 ha	1,500 ha	1,000 ha	750 ha	300 ha

savanna" can be as low as 225', which is around 700 trees per hectare, or the spacing equivalent of 4 m × 4 m (the density highlighted in Table 6.2). More importantly, the area needed is not large in relation to the area of the denser forest patches that one might expect to find in a village territory of the forest–savanna transition zone (i.e. peri-village forest islands, gallery forests, mature fallows), or the Guinean savannas. Equally, it is not large in relation to the area felled annually by a village within a sustainable fallow cycle. Indeed, it would be reasonable to suspect, according to Goucher's figures, that the fuel for smelting was merely a byproduct of farming. Thus, while staggering at one level – felling, splitting and making charcoal out of 300,000 trees is no mean feat – the area required to produce this wood is counter-intuitively rather small. Admittedly this fuel demand would be extra, on top of existing fuel consumption, but it is nevertheless quite a small extra.

In further support of her case, Goucher (1981) argues that fuel shortages helped transform the technology for iron production, implying that necessity was the mother of invention. Studies of slag, she asserts, indicate a change in charcoal fuel between the fourteenth and eighteenth centuries. She contends that a new technology was introduced which produced the desired characteristic of high heat (i.e. pre-heating) from less fuel, and that this was a 'fuel-conscious' adaptation' which may also have been a response to the 'ecological degradation and deforestation'. But it may not. Once again no clear evidence is provided. A counter-argument is that the labour required for producing charcoal, and not wood supply itself, may help interpret why fuel-saving technology was preferred. Would it not be in every smelter's interest to save time and energy (money) by using a superior, charcoal-efficient technology? Charcoal is never 'free', even when wood is abundant. Goucher does note in support of her case (citing Duncan 1847: 132) that there was frequent uncertainty about obtaining charcoal for the furnace, but one could retort that this problem of supply is just as likely to have been uncertainty about gaining charcoal of adequate species quality, or at the right labour or financial cost, as it is to have been due to the general lack of woodland. Goucher's argument that improved technology responded to 'degradation' is, once again, forced.

She goes on to argue that 'when these technical advances could not overcome the ecological challenge, the only viable response by African metalworking industries was an increased reliance on the industrial products of Europe'. Such a conclusion, we think, cannot be sustained from the data she presents. Other reasons for the decline of African smelting need to be considered.

Goucher (1981) argues that fuel shortage is indicated by evidence for long-distance trade in appropriate charcoal fuel from Asante to Bassar (and of iron in the reverse direction). She infers that smelting in Bassar may have been partly responsible for northern Ghana's dewooding as well. Rather than going through these calculations to estimate the area required to furnish the 'staggering' number of trees, Goucher makes several anecdotal remarks in support of her case. First she commits the error of transposing European experience onto African conditions. 'In medieval Europe,' she writes, citing Gimpel, 'the provision of fuel for a single smelting furnace could level the forest for a radius of one kilometre in a mere 40 days of operation' (Gimpel 1976, cited in Goucher 1981: 182). Even assuming that European charcoal smelting was not species specific, this seems something of an exaggeration. A circle of 1 km radius has an area of 314 ha, which, with trees spaced at 4 m as might be the case in mature European 'forest', would total almost 200,000 trees. If each medieval person converted a tree a day into charcoal (a herculean task with nothing but an axe), this would require about 5,000 people working full time over those 40 days within that 1 km radius. Does the timescale of 40 days indicate that this is a mere apocryphal story? Second, she asserts – without evidence – that the 'thousands of trees exploited by one iron smelter probably could not have been replaced during his lifetime'.[9] But in the line above she suggests that trees of the African savannas regenerate within perhaps 20 years.

Finally – and this is why we have dwelled for so long on Goucher's thesis – she argues that in West Africa the area of most intense iron production includes the region now considered by some scientists to be 'derived savanna'; that is, she says, 'a type of savanna created by human intervention'. Smelting-driven deforestation has been used uncritically in several influential studies reflecting on West African vegetation history (e.g. Nyerges 1988; Brooks 1986, 1993).

She notes from the early Portuguese accounts that certain areas were producing iron such as the Fouta Djallon and links this with the assertion that such areas, now grasslands, were 'originally' forested (citing Church 1968 as evidence). From this she infers that it was ironworking that did the damage. The author gives only two other sources of evidence for savannisation in the region: the analysis of Dorward and Payne (1975) in Sierra Leone, a critique of which is given in Chapter 7; and a naive reading of the rationale behind early colonial policy. She writes that 'the French were forced to enact legislative protection against further over-exploitation of the environment in Senegal and to provide for systematic replanting' (Goucher 1981). In other words, she assumes that if the French colonial regime had a policy, there must have been a problem. This is the same heroic assumption she made earlier, in relation to German policy in Togo, and it is open to the same critique which we have already made.

As we have noted, calculation of areas of forest needed for smelting at the levels indicated by Goucher suggests that iron smelting would have hardly contributed to deforestation in this part of West Africa (but see Warnier and Fowler 1979 for Cameroon). Indeed, so-called 'derived savanna' usually has forest patches within the landscape (forest islands, gallery forests, forest fallows) of areas in the range of those needed to fuel iron extraction at these rates. Forests/trees in this warm, sunny and well-watered region have some of the highest growth rates recorded. Smelting operations were, in any case, often located near to forest patches (see Anderson 1870). One can conclude that if the data given by Goucher are representative, this region could easily accommodate iron smelting sustainably at the recorded level. Had smelters also managed their wood supply, this would be all the more so – but this issue has not been researched.

Iron smelting in Togo

While Goucher (1988) applies her argument to deforestation in Togo, she does not systematically use data from the country. Nevertheless, more quantitative data concerning iron production in Togo now exist and it is worth examining them in more detail in the light of Goucher's arguments and our critique of them.

That iron smelting was carried out on a large scale in Togo is not in doubt. Smelting in Togo centred on the Bassari region, an area of about 100,000 ha in the centre of the country, some distance from the forest zone. Kuevi uses colonial reports to estimate that iron production in the Bassari region might have been in the order of 200 tonnes/year (Kuevi 1975: 43). This kind of production level is also borne out in the detailed archaeological analysis of de Barros (1986, 1988). He estimates that: 'Total Bassar iron production since the industry's inception in the late first millennium AD has been conservatively estimated at between 14,000 and 32,000 tonnes [c. 16–36 million Bassar hoe blades]', placing Bassar among the top iron-producing centres of Africa (de Barros 1988: 95–6). He also argued that the original savanna woodland vegetation had been reduced to a basically savanna landscape by extensive charcoal making for the iron industry and by increasingly short periods of agricultural fallow (de Barros 1988: 92).

Using the same figures and assumptions as in Goucher's analysis, and our critique of it, we can calculate how much land would be needed to supply this much charcoal. This is summarised in Table 6.3.

While Table 6.3 suggests that between 100,000 and 250,000 ha of land would have been cleared of trees, this is the area for over 1,000 years, over which time trees would have regenerated perhaps thirty times. From these calculations, it emerges that for each tonne of iron produced, about 8 ha of land would need to be used for charcoal; i.e. 113,659 ha divided by 14,000 tonnes of iron. Yet as de Barros provides more detail concerning annual iron production during different historical phases of iron production, one can go on to calculate more realistically the area of forest required annually to supply the iron industry during its different phases – data presented in Table 6.4.

From Table 6.4, it appears that even to maintain the upper limit of

Table 6.3 Land area needed to produce fuel for 1,000 years of the Bassar iron industry

Charcoal needed to produce 1 tonne of iron (c. 1810)	*Wood needed to produce 1 tonne of iron @ 10%*	*Wood needed to produce 1,000 year production of iron (lower limit 14,000 t; upper limit 32,000 t)*	*Trees needed to supply wood (@ 27 kg/tree)*	*Area required to supply trees for 1,000 year production (@ 625 trees/ha)*	*Area required to produce this wood sustainably (assuming 33-year regeneration to only 27 kg/tree)*
13.7 t	137 t	1,918,000 t	71,037,000	113,659 ha	3,788 ha
		4,384,000 t	162,370,000	256,000 ha	8,533 ha

Note
These figures derives from Rosillo-Calle, F. (pers. comm. to Gerald Leach, 2 June 1995).

Table 6.4 Estimated area of woodland needed annually to supply the Bassar iron industry during different historical phases of output

Period	*Tonnes/annum*	*Area of woodland for fuel/year (ha)*	*Area needed for sustainable production (@ 30-year regeneration) (ha)*
1 Pre-AD 1300			
Upper limit	<7	56	<1,680
Lower limit	<20	160	<4,800
2 1300–1550			
Upper limit	7	56	1,680
Lower limit	20	160	4,800
3 1550–1800			
Upper limit	28	224	6,720
Lower limit	81	648	19,440
4 1800–1905			
Upper limit	60	480	14,400
Lower limit	135	1,080	32,400

production at the peak of the iron-producing era, an area of only about 32,000 ha would be capable of producing the required wood sustainably (a circle with a radius of 10 km). The area of Bassar is perhaps three times this, at 100,000 ha. Furthermore, only 1,100 ha of land would be needed annually, which must be much less than the area cleared annually by Bassari farmers for their fields.

There are many incorrect assumptions in this sort of analysis, which is conducted only to gain an idea of orders of magnitude. Above all, not all trees are used for charcoal. But to the extent that more area is required to supply the correct species, so the land is not dewooded, but possibly only depleted in important trees.

Oral traditions suggest that at the era's height in 1900, iron produced in Bassar was traded across Togo, eastern Ghana and western Benin, servicing perhaps 600,000 people over 10 m ha. Thus one can also argue that even if these calculations are wrong, and especially if vegetation degradation occurred in particularly intensive smelting areas and particularly intensive smelting times, in an environment that had fewer trees than envisaged, the impact of this smelting on regional vegetation would not have been large – even if it was significant to Bassar. In short, we argue that the impact of indigenous pre-colonial iron smelting on the environment in West Africa, and in particular in Togo, has been much exaggerated.

Similar critiques have recently been made of the exaggerated effects of domestic woodfuel demand in instigating forest clearance (e.g. G. Leach and Mearns 1988; Dewees 1989). The image of a widening 'woodfuel gap', it is argued, depends on several faulty assumptions, compounded by measurement errors in calculating woodfuel supply and demand. Deforestation, the critiques argue, is rarely a direct result of fuel demand (as the gap theorists assume) but more often of land clearance for agriculture and other purposes, with woodfuel frequently a byproduct of this. In assuming that fuel demand translates directly into tree loss, the woodfuel gap model also ignores both the widespread use of shrubs, bushes and crop residues for fuel, and the ways that consumers adapt their consumption rates in response to perceived scarcity.

Conclusion

Although few authors consider there to have been major deforestation in Togo during the twentieth century, forest cover change is still framed in terms of deforestation. While the process of deforestation has been back-dated to earlier periods and, as we have seen, to other causes, the same unrelenting logic of deforestation remains strong. There is, however, little evidence concerning pre-colonial vegetation cover in Togo. Furthermore, there is evidence that farming can install as well as degrade forest.

The Togo case exemplifies in a detailed way how forest cover analysis can be seductively linked to analysis of economic and technical change, such that the two become mutually dependent. By questioning one aspect – and here we have strongly questioned the thesis of a woodfuel gap based on iron smelting – the entire analytical edifice can begin to appear as a house of cards. A similar endeavour is presented in the next chapter, where analysis of linked social, technical and epidemiological change has been influential in forming dominant views of deforestation in Sierra Leone.

We would argue that as a result of the pathbreaking study of Guelly *et al.* – in keeping with similar works in the transition zone – the history of vegetation change in Togo could be reframed completely. Inconsistencies in forest cover data might also be taken to suggest this. Analysis of vegetation dynamics in Togo, as in the other countries reviewed so far, needs to be tied more closely into analysis both of historical vegetation management in different economic eras and of demographic shifts linked to political events as much as to the logic of population expansion.

Notes

1 This refers to German Togoland, half of which was transferred to British Mandate after World War I and is now part of Ghana.

2 We shall forever remain dumbfounded by Gayibor's (1988) reply to Blanc-Pamard and Peltre (1987), in which he uses a quotation from Zoller (1885) to suggest the presence of forest on the coastal strip of Togo. It does not.

3 Despite being so far north, one should note that this area of Togo receives c. 1400 mm of rainfall.

4 She writes: 'Deforestation produced by environmental exploitation (particularly reliance on charcoal fuels) must be taken into consideration.'

5 One needs, according to Goucher, 'a slow-burning, dense, hard wood, usually with high alkali and silica contents and a granular structure' (Goucher 1981).

6 While Goucher suggests that certain species are 'used' to make charcoal for smelting, she does not specify whether these are essential or merely preferred species.

7 A tree of 0.2 m^3 at 1,310 kg/m^3 might weigh ten times Goucher's estimate, 262 kg.

8 Note that today the World Bank assumes that charcoal weighs 12 per cent of its parent wood. In Tanzania, 18 per cent is commonly achieved. The most efficient conversion could be c. 25–30 per cent (Gerald Leach, pers. comm.). We should note that we are using Goucher's own calculations here, which assume that to produce only 1 kg of slag, it takes the charcoal from 1.5 trees.

9 In fact she cites Richards (1973: 65), but this does not seem to give the relevant information.

7

SIERRA LEONE

Introduction

Sierra Leone has not escaped exaggerated estimates of recent deforestation. This has been attributed not just to farming and land use practices, but also to the timber trade, thus highlighting not just the local actors identified in many deforestation analyses but also the supposed role of external ones. A major aim of this chapter is to show how views of deforestation have supported, and been supported by, a particularly detailed analysis of socio-economic and technical change hingeing on the supposed ecological and epidemiological consequences of the timber trade. This is the last of our country case studies, and representation of Sierra Leone's forest history has relied on a complex interaction of all the deductive strategies which build images of forest pasts. In consequence, our critique draws on all the critical strategies developed in earlier chapters and is thus well situated at the end.

A number of analyses hold the country to have lost major forest tracts during the twentieth century. Perhaps the most extreme comes from Myers, who wrote that 'as much as 5,000,000 ha may still have featured little disturbed forest as recently as the end of World War II. It is a measure of the pervasive impact of human activities that the amount of primary moist forest now believed to remain is officially stated to be no more than 290,000 ha' (1980: 164). Nyerges compared two vegetation maps drawn twenty-four years apart (Keay 1959a; White 1983) and deduced 'a significant change for the westernmost extension of the forest in Sierra Leone, which has been entirely converted to savanna over the years since World War II' (Nyerges 1987: 328). That deforestation has been recent and large is also inferred by Sayer *et al.*, who suggest that '50 per cent of the country has climatic conditions suitable for tropical rainforest, but less than 5 per cent is still covered with mature, dryland closed forest. Deforestation is mainly a result of the rapidly increasing human population requiring more agricultural land and fuelwood' (1992: 244). FAO (1993) estimated that 123,000 ha of 'tropical rainforest' were cleared between 1980 and 1990.

While it is certain that Sierra Leone currently has a relatively low proportion of mature forest,[1] the question arises as to when, if ever, it had much more. As Richards (1994, 1996) has frequently affirmed, figures suggesting post-war forest devastation in Sierra Leone have little in common with earlier estimates of forest cover.

The foresters Unwin and Lane-Poole visited the country in 1909 and 1911 respectively to assess Sierra Leone's forest resources for the British colonial service and to establish a Forest Department. Unwin suggested that: 'Probably in the earliest times the whole territory was covered with some kind of arborescent growth varying from open savanna and deciduous forest to close impenetrable evergreen rain forest. Now scarcely 1% of this forest remains' (Unwin 1909). Lane-Poole put it the other way round: 'Ninety-nine per cent of the rain forests have been destroyed by the natives in their wasteful method of farming' (Lane-Poole 1911: 4).

Sierra Leone's report to the first British Empire Forestry Conference stated that the 'primeval forest' is now 'almost disappearing'. 'With the advent of the Pax Britannica . . . the native peoples have had a greater opportunity of multiplying and of devoting themselves to agriculture. Within a lifetime they have cleared up to about 90 per cent of the virgin forests by their system of shifting cultivation' (Thomas 1924: 5). 'There are evidences that at no remote period the whole country was covered with dense high forest. At present the area covered by high forest is estimated to be in the neighbourhood of 3 per cent of the whole; and this is likely to be a generous estimate' (1924: 7).

Other descriptions by early travellers in Sierra Leone also indicate the lack of extensive forest tracts early this century. For example, in the 1920s the travel writer Migeod wrote that:

> In former times there was a great forest area south of the railway line that runs due east from Freetown direct to the Liberian border. *Now* there is scarcely a vestige left until the Liberian border is actually reached.
>
> (Migeod 1926: 331, our emphasis)

In 1930, Meniaud suggested that Sierra Leone had no more than 300,000 ha of forest (1933). Later, in 1937, official estimates had risen to suggest that 6 per cent of Sierra Leone remained forested (Government of Sierra Leone 1937: 9). It is figures such as these which have led other recent analysts to be more cautious in affirming recent deforestation. Gornitz and NASA (1985) and Parren and de Graaf (1995), who cite them, for example, found such figures in Zon and Sparhawk (1923), who suggested that there 'now remains scarcely more than 87,200 ha, or 1.0 per cent' (1923: 834). Indeed, these early figures appear to describe a much lower forest cover in the country (1–3 per cent of Sierra Leone's 7,162,000 ha) than recent estimates suggest exists there today.

Thus many present-day assessments, more informed of conditions around 1900, date Sierra Leone's deforestation not to the later part of the twentieth century but to the early and mid-nineteenth century, as did early colonial analysts. Here we examine the arguments forwarded for deforestation during this nineteenth-century period; arguments resting on the environmental impact of timber exports. These have been accredited such significance by two influential recent studies (Dorward and Payne 1975; Millington 1985) that Sierra Leone has come to be treated as a nineteenth-century forerunner to the twentieth-century deforestation elsewhere in West Africa (e.g. Gornitz and NASA 1985; Parren and de Graaf 1995). We ask whether the nineteenth-century timber trade, rudimentary as it was then, could really have had such an environmental impact, and whether there was such extensive forest there in the first place. We then go on to examine the evidence on which analysts have written the history of Sierra Leone's vegetation as a one-way loss of an original forest cover, whether this century, last century or before (see Map 7.1).

Nineteenth-century deforestation and the timber trade

Sierra Leone became an exporter of timber from the 1820s. That this trade had a major impact on Sierra Leone's vegetation was suggested by Migeod in the 1920s:

> As late as seventy years ago timber was a principal article of export from all that region well furnished with waterways in the shape of the Scarcies and Rokelle rivers with their branches. Now one might walk up and down the country for days looking for a fair sized tree. Not only have the big trees gone, but the younger ones are no older than ten years growth at a maximum, at which period they are burnt to prepare the land for a farm. All one can find are inferior trees along the river banks.
>
> (Migeod 1926: 331)

More recently, the importance of this timber-led deforestation has been underlined by Dorward and Payne, who, in an influential and widely cited study informing the later work of several authors (Goucher 1981; Gornitz and NASA 1985; Nyerges 1987, 1988; Parren and de Graaf 1995), linked it to veterinary epidemiology. They argue that the epidemic horse disease of 1856–70 in the capital Freetown was caused by the appearance of tsetse fly and the disease, trypanosomiasis, which it carries. This occurred, they argue, because of deforestation. 'Some change in the prevailing ecological conditions must have occurred which favoured the expansion and increase of the

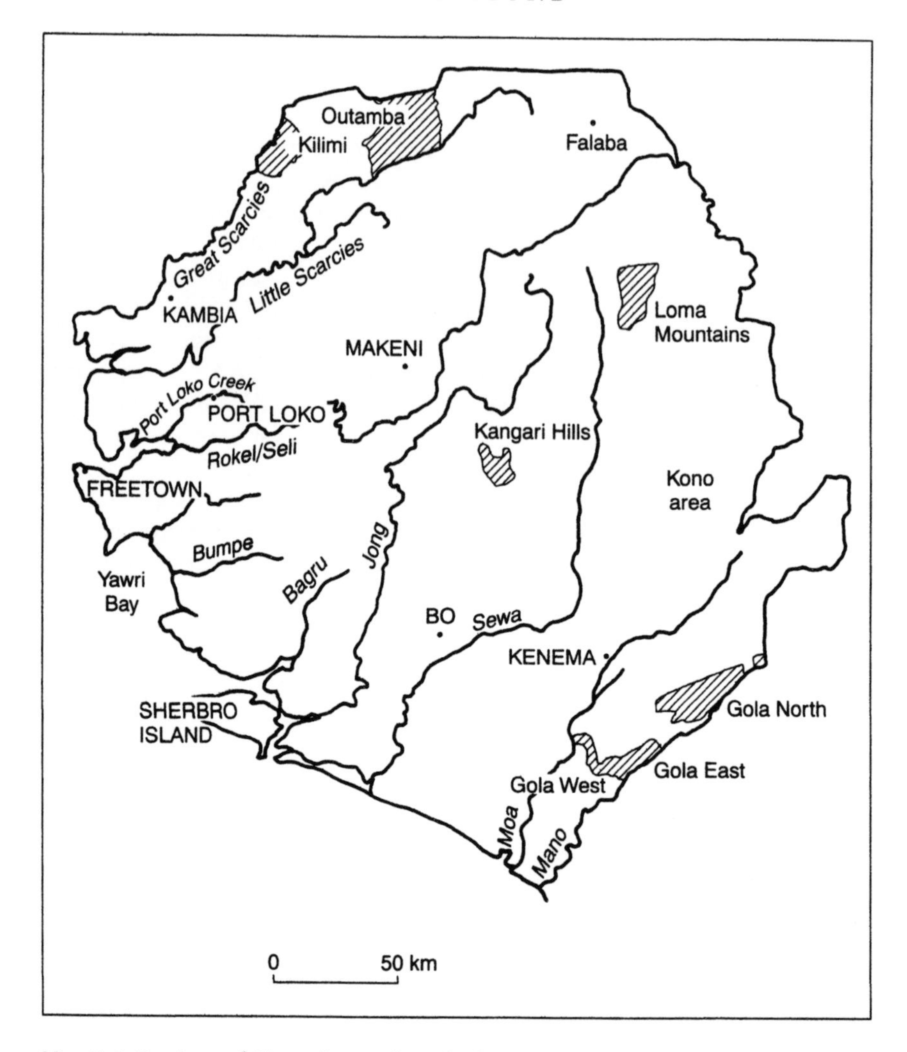

Map 7.1 Regions of Sierra Leone described in the text.

tsetse fly population and brought it into contact with the horses on the peninsula' (Dorward and Payne 1975: 246). Their argument is that 'much of Sierra Leone was originally covered with tropical high forest in which the tsetse flies mainly associated with typanomsomiasis do not occur' (1975: 246). They cite Nash (1948) and Buxton (1955) to assert that tsetse does not inhabit high forest. Thus they argue that with an extensive hinterland cover of high forest, the Freetown peninsula would have been isolated from tsetse-borne infections since the forest would act as a barrier to the expansion of the tsetse fly population either in terms of radiating from riverine foci (as in the case of *G. plapalis*) or incursion from the savanna (as with *G. morsitans*, a common vector of *Trypanosomiasis brucei*).

Dorward and Payne purport to describe extensive evidence of deforestation. They argue that forest in the immediate vicinity of Freetown had been felled early in the history of settlement, citing the botanist and collector Afzelius (1792), who himself noted the prior 'deforestation' and needs for conservation. The authors then go on to document deforestation in the nascent colony's immediate hinterland, linked to timber felling over three distinguishable periods: 1816–30, when first-grade timber was extracted; 1830–50, when small-scale merchants cleared the remaining timber, and, overlapping with the latter, from the 1840s, when farmers colonised the logged-over land.

From 1816 to 1820, three logging merchants set up just outside the nascent colony's jurisdiction, mostly to supply timber to the British Admiralty shipbuilders. Initially these merchants acquired timber from the banks of the Rokel river (around Rokel village) and around Port Loko and other creeks. The wood taken was 'African Teak' or 'No.1',[2] which seems to have referred not only to *Oldfieldia africana*, but also to a range of good-quality 'No. 1' timber species, including *Lophira alata* and *Milicia excelsa* (see Dalziel 1937: 156). Dorward and Payne cite one resident, McCormack, to suggest that there were already problems of supply as early as the mid-1820s:

> 'I do not think the timber denominated African teake, or No. 1, will be procurable in sufficient quantity for more than seven or eight years in this river [the Rokel]; from the circumstances of the distance, the natives will have to haul it out of the woods, and from the state of the country, it is impossible to use carriages of any description; and I do not think the natives would be disposed to make roads, at least at present. There are, however, a number of other valuable woods, fully equal in my opinion to No. 1 which will last for many years to come.'
>
> (cited in Dorward and Payne 1975: 248)

Yet while this citation alludes to supply problems, it also suggests that timber extraction was highly species selective; less 'deforestation' than selective extraction from navigable streamsides. It does not support the idea of vast tracts laid waste by the timber trade, as Dorward and Payne seem to infer. Later, by 1829, new sawpits had opened on the Scarcies and neighbouring rivers, and by 1831 the principal merchant McCormack had moved from the Rokel river area to re-establish his operation at Gbinti at the mouth of the Melakori river. This signalled a departure of the big exporters from the Port Loko and Rokel areas, and small operators, many of them liberated slaves from Freetown, moved in to clear out what timber remained.[3]

Dorward and Payne go on to argue that the original clearance of the forest along the river and creeks in the northern areas paved the way for agricultural development (as is the case in many twentieth-century models of deforestation), giving as an example McCormack's sale in 1830 of his timber factory at Gbinti and on Kikonke Island in the Scarcies to the wealthy trader and entrepreneur Charles Heddle, who then proceeded to use the land for groundnut cultivation.

Dorward and Payne argue that the shift in location from the 1830s was necessary as the northern rivers had become less profitable than the more southern Ribi river, the Bumpe chiefdom on Yawri Bay, and Bagru in Sherbro. The inference is that the northern rivers were being worked out. A better inference, however, may be that as the Sierra Leone colony was gaining strength, these other – perhaps cheaper – supplies of timber were becoming accessible to its traders through new political deals. In any case, if the northern rivers were worked out, producing scarcity, this may not be because timber had already been felled on such a huge scale, but because there was not much timber about in the first place. One way to approach this latter question is to examine the quantity of timber exported up to that time, and to calculate the area of forest required to produce it. Strangely, none of the authors arguing for the importance of nineteenth-century timber felling have made such estimations.

At that time, timber was conveniently measured in 'loads', which is the customary quantity which could be carried in a cart or wagon, and for timber this was conventionally 50 cu. ft. Export data for Sierra Leone are given in Table 7.1. Generally, equatorial forest in West Africa would supply between 15 and 50 cubic metres (see Chevalier 1909c for Ivorian forests). Houdaille (1900) estimated a production of 40 cu. m/ha of timber suitable – or better – for ship building.[4] Given the value of the wood, and that it was hand cut, we can assume that there was no great wastage and that these commercial figures are reasonable for the type of tropical forest which Sierra Leone's analysts believe to have disappeared. Assuming this 40 m^3/ha yield, the area from which timber was taken between 1827 and 1843 at the height of the felling can be estimated at 9,258 ha (i.e. less than 10 km × 10 km), averaging 544 ha/year. In total, one can calculate that timber was taken from about 15,000 ha (i.e. an area of c. 12 km × 12 km), or less than 400 ha/year (1820–60). Up to the time when the timber trade moved south, from the Scarcies and Port Loko, timber had been extracted from perhaps some 5,000 ha, an area of c. 7 km × 7 km. That timber was removed from such an area does not, of course, even suggest that this area was 'deforested', first because mature forest regenerates rapidly after felling, and second because the felling was so selective. Despite the crude nature of such calculations, it is reasonable to argue that the rate and extent of nineteenth-century timber felling were almost insignificant relative to the area of Sierra Leone. If there was a shortage of timber during the period of

Table 7.1 Forestry exports from Sierra Leone, indicating the rate of deforestation per annum (1827–60)

Year	*Teak-wood exports (loads)**	*Exports of forest produce (Unwin 1909)*	*Cubic feet (using 1841 data)*	*Cubic metres*	*Ha forest @ 40 m^3/ha*
1827	15,625	10,742	781,250	22,321	558
1828	11,892	1,114	594,600	16,988	424
1829	15,992		796,100	22,745	568
1830	19,363		968,150	27,661	691
1831	23,650	18,983	1,182,500	33,785	845
1832	15,119	24,048	759,950	21,712	542
1833	13,304	1,771	665,200	19,005	475
1834	13,192	16,951	659,600	18,845	471
1835	14,013	9,302	700,650	20,018	500
1836	13,243		662,150	18,918	472
1837	22,655		1,132,750	32,364	809
1838	10,628		531,400	15,182	379
1839	10,622		531,100	15,174	379
1840	11,601		580,050	16,572	414
1841	12,616		630,800	18,022	450
1842	13,243		662,150	18,918	472
1843	22,655		1,132,750	32,364	809
1844	6,000*		300,000	8,571	214
1845	7,000*		350,000	10,000	250
1846	8,000*		400,000	11,428	285
1847	7,500*		375,000	10,714	267
1848	7,250*		362,500	10,357	258
1849	7,000*		350,000	10,000	250
1850	3,000*		150,000	4,285	107
1851	6,500*		325,000	9,285	232
1852	8,000*		400,000	11,428	285
1853	8,000*		400,000	11,428	285
1854	6,500*		325,000	9,285	232
1855	5,500*		275,000	7,857	196
1856	3,000*		150,000	4,285	107
1857	5,000*		250,000	7,142	178
1858	7,000*		350,000	10,000	250
1859	4,000*		200,000	5,714	142
1860	2,000*		100,000	2,857	71

Note
* Data from 1827–41 are derived from the appendices in the proceedings of the 1841 select committee on Conditions in West Africa. Data for 1844–60 are to be found in Sierra Leone's Bluebooks, but those presented here have merely been roughly extracted from Dorward and Payne's own (1975) graphic presentation.

nineteenth-century export, it seems more likely to be because there was not much in the first place than because the felling was so rapacious. When it is stated that 'timber has to be dragged several miles to a river bank' this seems less likely to be because the intervening land had been deforested (as

Dorward and Payne infer) than because the intervening land did not hold forest of the desired quality, or forest at all.

While the timber trade probably did not have a major environmental impact, its economic importance to the early colony cannot be so minimised. As one contemporary analyst put it, 'But for the trade in timber, one which by the destruction of forests within reach of water carriage (and already the trees felled are often dragged many miles to the river) must exhaust itself, the commerce of the place must have sunk into insignificance' (Mr Forster 1842, evidence presented to the Select Committee on the Slave Trade and Sierra Leone, British Government). It may have been in anger at economic expropriation, more than for ecological reasons, that timber extraction sometimes incited local resistance. When, for example, in 1852 the men's secret society of the Rosolo creek region put a prohibition or 'poro' on tree extraction, it is estimated that £16,000 worth of felled timber was left idle (Fyfe 1962).

Dorward and Payne go on to suggest that by the late 1860s the timber trade from the southern rivers (Ribi and Bagru) was also declining, owing partly to the exhaustion of profitable reserves and partly to competition from the growing palm kernel trade. They minimise how by then timber prices had collapsed, along with Admiralty demand after ship building switched from wood to iron, making the role of supply problems in the decline highly questionable. The authors also suggest that the palm kernel trade 'followed in the wake of the clearance' for timber (1975: 249). But how much had been cleared for the timber trade – perhaps only c. 2,000 ha over the 1850s. Thus, the inference suggested to the reader of a landscape denuded of forest owing to timber felling and later agricultural clearance and replacement with palms must be wholly incorrect.

The volume of timber extracted and the area extracted from can reasonably be accounted for by the extensive network of streamside riverine forest found in this part of Sierra Leone. As Scott Elliot summarised in 1893, 'it is almost an invariable rule that every river and stream possesses a belt of forest from 50–100 yards wide on each bank. The trees in these river belts are sometimes of considerable height. These belts of timber continue up to various streams and tributaries almost to their source, becoming, however, much shorter and smaller towards the end' (1893: 33). Only 2,000 km of such forest, not susceptible to savannisation, would have been needed to provide the timber involved in the nineteenth-century trade, a length easily accommodated within the immediate hinterland of Freetown, given the multiply dividing streams in the catchments.

Dorward and Payne nuance their argument, suggesting that forest cleared for timber was largely from areas near rivers because fellers were dependent upon the rivers for the collection and transport of logs, arguing that the intervening lands were subsequently cleared by farming. Yet it could again plausibly be argued that in the mid-nineteenth century riversides were the principal, if not the only, areas where suitable high forest could be found. If there were large tracts of intervening forest, when were they cleared by farmers? Dorward and Payne suggest that this was gradually from the late nineteenth century, and most dramatically from 1912. Thus, while arguing that major deforestation in Sierra Leone was timber-led, they neverthless fall back on twentieth-century farming to account for 'the final clearance of the forest and its replacement with farmland and derived savanna' (Dorward and Payne 1975: 251). Citing Cole (1968: 12), they argue that: 'once "formal" agriculture had been established over most of the country in the period after 1912, the remainder of this high forest was converted into low bush within 20 years.' This argument can be rejected for many reasons. First, the observations of foresters cited above made just prior to the 1912 report suggest that Sierra Leone had almost no forest at this time. Second, neither Dorward and Payne nor Cole suggest what happened in 1912 to render Sierra Leonean farmers' agriculture 'formal'. They seem – mistakenly and naively – to be suggesting that rainfed farming on the uplands was new to Sierra Leone. Third, the picture of the late nineteenth century as a time when farmers were gradually moving into tracts of previously uninhabited forest during a period of economic development and population increase is quite the contrary of the period's documented history. The late nineteenth century was a time of major upheaval, warfare and associated economic slump in many regions of Sierra Leone, with decimation of previously greater farming populations. For example, of the areas inland from the south-eastern rivers documented by Dorward and Payne, Menzies wrote in 1870 how 'ceaseless wars have laid waste the country, and their traces may be found in the numerous sites of ruined towns' (Church Missionary Intelligence 1870: 85).

Millington is a second author to assert, independently it would seem, that there was major timber-led deforestation in the nineteenth century. He cites Cole (1968) to assert that the demand for timber was so great that 'by 1860 all the wood along the Little and Great Scarcies, Port Loko Creek and Rokel river and a belt of land between Waterloo and Bumpe was cleared of timber' (Millington 1985: 34). Cole had suggested that 'the supply of timber near the coast was exhausted by 1840, and exploitation progressed inland through Waterloo, Mabang and Bumpe chiefdom', but by 1860 the timber trade began dwindling for lack of adequate supply (1968: 3; see critique of this latter position above). But Millington goes on to exaggerate this area. He notes that 'the zones of economic production in the mid-eighteenth century [sic; he means nineteenth century] and the areas of deforestation coincide quite closely', and deduces that 'it is probable that the farmers moved into the cut areas after the departure of the woodcutters to grow their own crops for ease of clearance' (1985: 34–5). On an accompanying map purporting to show areas of nineteenth-century deforestation, Millington does not merely indicate the riversides and areas that Cole describes, but incorporates entire sweeps of territory (see also Millington 1987) totalling about 750,000 ha (rather than the 15,000 ha which was needed to supply the timber – less than 2 per cent of Millington's estimate).

In short, the argument for the importance of nineteenth-century deforestation centred on the timber industry in Sierra Leone is, we suggest, invalid. A different reason will have to be sought for the outbreak of trypanosomiasis among the horse population in the 1850s. Nevertheless, evidence for the nature and extent of Sierra Leone's deforestation does not rest only on analysis of the timber trade, or on the sources used by Dorward and Payne and Millington. We now go on to examine other types of evidence and assumption on which analyses of the changing nature of Sierra Leone's vegetation have relied.

Questioning the basis of deforestation analysis

Generally ecologists have interpreted Sierra Leone's vegetation in the context of the unilineal degradation of climax vegetation communities. Many authors suggest that for large parts of Sierra Leone the climax is dense evergreen or semi-deciduous forest. In particular, the regions today dominated by secondary forest thicket or farmbush are assumed once to have been covered by original forest. Equally, many regions today dominated by savannas are interpreted as one stage further in the landscape's degradation from forest. Such assumptions have dominated analysis of Sierra Leone from the earliest. Thus Lane-Poole in 1911, after documenting successive stages of degradation from forest to savanna due to farming, shortening fallows and fire, concluded that:

> In a country like this, where grasslands have already an advantage over woodlands, if the balance of nature is upset by the felling over a large area of all ligneous growth, the woodlands are entirely at the mercy of the grasslands. A conversion is the result; the rain forests become savanna forests and these in turn become pure savanna.
>
> (Lane-Poole 1911: 5)

As elsewhere in West Africa, these assertions have generally rested not on historical data, but on deductions of vegetational pasts based on present-day vegetation observations. This approach has been recently exemplified by Sayer *et al.* who suggest that across Sierra Leone 'land formerly covered in forest can often be identified by the presence of oil palms, which have economic value so have been retained by the farmers. Now only a very limited area of the country supports climax vegetation. However, small patches of closed forest (generally under five hectares) are often maintained near settlements as sacred sites for traditional religious uses' (1992: 244).

The assumption that forest patches are relics of original forest cover has long dominated analysis of Sierra Leone's savannas. In his *Vegetation of Sierra Leone*, Cole asserts that forest outliers in savanna regions probably originated as remnant forests; islands of original vegetation in 'a sea of grass or derived savanna' (1968: 81). Lane-Poole (1911: 6) described, as 'part of the remaining 1% of Sierra Leone's forest . . . a piece of forest is preserved around each town, a part of which is used as Porro bush' (i.e. as secluded sites for the activities of secret societies).

The critiques of this reasoning already discussed for Benin, Togo and Côte d'Ivoire are equally pertinent to Sierra Leone. As early travellers noted, forest outliers are generally found on existing or old village sites. The Liberian intellectual Blyden (1872), for example, noted such forests around all major Limba towns, which were 'always built on difficult and scarcely accessible highlands and protected by the cover of high forests'.[5] That such forest patches in savanna may partly have originated in fortifications, as our own research across the border in Guinea demonstrated (Fairhead and Leach 1996a), was noted by Migeod during his travels around Sierra Leone in the 1920s:

> It is no uncommon thing to see a small forest round towns and villages, when there is none surviving anywhere else. Chief among the trees is the kola . . . The Bombaces rear themselves above all the others covering much ground with their enormous buttresses. The origin of thick timber growth round a town was defensive purposes. The old stockades have taken root, and one may trace the lines of them in the big trees at the present day.
>
> (Migeod 1926: 334)

When Blyden visited Falaba in north-east Sierra Leone in 1872 he imagined that the silk-cotton trees which dominate so many of Sierra Leone's forest outliers were natural. Falaba, he wrote, is 'surrounded by a *natural* stockade of over five hundred huge trees – one hundred and ninety of which are very old, and enormous silk cotton trees. One of the gates of the town, of which there are seven, is ingeniously cut through the trunk of one of the largest trees' (Blyden 1872: 12, our emphasis). Yet the explorer Major Laing, who had visited the town fifty years earlier, noted that: 'The whole settlement was surrounded by a thick stockade of hardwood' and that the stockade had 'taken root in many places and grown into large trees among the branches of which the Soolimas station themselves' (Laing 1825: 352–3).

A critique of vegetation analysis in northern Sierra Leone should not stop at reinterpreting forest islands. It also puts into doubt the validity of treating secondary thicket always as degraded forest. In the Kuranko areas near the Loma Mountains, for instance, farmers claim that that 'their cultivation could influence the direction of succession towards forest, particularly from savanna. . . . Farmers in the savanna areas specifically protect some areas of the savanna from fire using fire breaks so that they can regenerate moist deciduous forest' (Pocknell and Annalay 1995: 19–20). Such deflection of vegetation succession today also appears to be associated with the cultivation of pigeon peas, which has the effect of suppressing the most flammable grasses. Further west, our own consultation of Susu farmers revealed their strong awareness of techniques to upgrade savannas to secondary forest thicket fallows using a combination of intensive grazing and organic matter incorporation, stressing the role of termite activity in this process.[6] These accounts of processes of vegetation change from savanna to farmbush undermine the near-universally held assumptions that farmbush indicates degraded forest and that savanna represents a further degradational stage.

Millington (1985) deduces the presence of degraded, as opposed to edaphic, grasslands by noting that areas of the country with the greatest proportion of upland grassland are also those with the greatest areas of short fallow bush regrowth (as opposed to longer fallows). He infers from this that the grasslands are the result of high cultivation pressure. Yet the observation that farmers upgrade savannas into short-rotation bush fallow undermines or even reverses this deduction. Where grasslands and short-rotation bush fallowing are found together, it may be that farmers have actively encouraged the grassland to develop into bush.

Many areas of central and southern Sierra Leone are now characterised solely by farmbush with patches of high forest, often around villages and shading tree crop plantations; patches which have commonly been taken as remnants of the region's 'original' forest vegetation. But even this deduction can be questioned on several counts. First, the significance of trees in fortification is not confined to the northern savanna regions. It is therefore possible that high forest patches owe their origins – or at least their particular quality – to defence structures established in already farmed and fallowed land. Siddle (1969) described how 'in the forested lowlands, every village was a "small fortress" maintaining several lines of defence' (see Little 1967: 27). Surrounding each village was a ring of high forest which camouflaged the village. While this may have been thick natural forest selectively preserved for the purpose, the fortress trees may have originated from stockades (see Alldridge 1901; Davies in Siddle 1969). As Harris (1866) described in Gallinas country, 'the fences are made of live sticks, planted about three inches apart, and which take root quickly; these have other sticks bound across them with a very strong and pliable vine . . . a second fence, of similar construction, but with the sticks nearer together, is placed about six feet within the first; and there is sometimes a third fence, but farther in the interior; where suitable wood is not available, walls of solid mud or clay are substituted for fences' (1866: 28). Many such walled towns exist in what is today the heart of Sierra Leone's forest zone (see Jones 1983b).

Second, where high forest patches overlie tree crops, the assumption that there was forest prior to tree crop planting can be questioned. Consideration of indigenous tree cropping strategies suggests that this may not always be the case. To the present, Mende farmers have at least two patterns of tree crop establishment (Leach 1994). While one of these does involve thinning mature forest, the second involves establishing cocoa on food crop land, where interplanted food crops give shade during the first 2–3 years. Selected tree seedlings are encouraged to grow up for shade and for their other uses. Thus a 'forest' with an understorey of tree crops can originate on land which might earlier have been under food crops and short fallows for decades.

Furthermore, plantations can be on the sites of old villages and farm camps, where the transformed soils not only favour good tree crop growth, but also influence the species and form of adventitious vegetation. In other words, the forest establishing on such sites may be indicative neither of the region's natural vegetation, nor of an unbroken forest past. In several instances, Mende farmers indicated to Leach how their ancestors previously inhabited the land now under their tree crop plantations, on occasion even pointing out still evident outlines of huts and garden mounds. That Mende single out and select past inhabited lands for their fertility is clear from the way in which finding stone sculptures in a field is taken as an indicator of field fertility and good farming fortunes. A further reason for suspecting many sites today under tree crops to have a long history of occupation and agricultural use is that the sites most favoured for tree crop planting nowadays are precisely the lower slopes and terraces (*bui*) of the landscape's catena, which (along with the neighbouring swamps) were the preferred sites of rice farming in the nineteenth century and before (Small 1953; Davies and Richards 1991; Richards 1996). For Mende, much tree crop plantation land represents a continuation of socially transformed space, rather than land hewn out of natural forest – contrasting absolutely with the way these dense patches of vegetation and shady canopies have been considered by many foresters.

High forest patches, whether in a landscape predominantly of savanna or of farmbush, may thus owe their origins not to selective preservation from deforestation wrought elsewhere, but to processes of active establishment and transformation by inhabitants. Somewhat ironically, then, the vegetation forms which forest ecologists select as indicative of least disturbance may in reality be those most disturbed. When coupled with analysis which suggests that land today in a bush fallow cycle may have been upgraded from savanna, or, as elsewhere in West Africa's forest zone, may have been savanna in earlier times, this possibility makes simplistic deductions about original vegetation based on present-day observations begin to look extremely suspect.

The question then arises as to how much of the country is, or has ever been, within the 'zone' where dense humid forest even might have existed. Of all West African countries, Sierra Leone presents the greatest ambiguity in the delimitation of its 'forest zone', with some authors placing the entire country within the humid forest zone, and others placing it completely within the Guinea savanna and transition zones (see Map 7.2).

Early vegetation maps and descriptions veered more towards the latter view. Thus Chevalier's original zonal map of West Africa (1912a) placed only the extreme south-east of the country in the forest zone, and Johnson (1906) did the same. As the colonial period progressed, however, redefinitions placed increasing areas of the country within the forest zone, suggesting that while they might not carry significant forest, they used to. The 1938 government forestry report, for example, redefined large parts of the north into a humid semi-deciduous forest zone, arguing that:

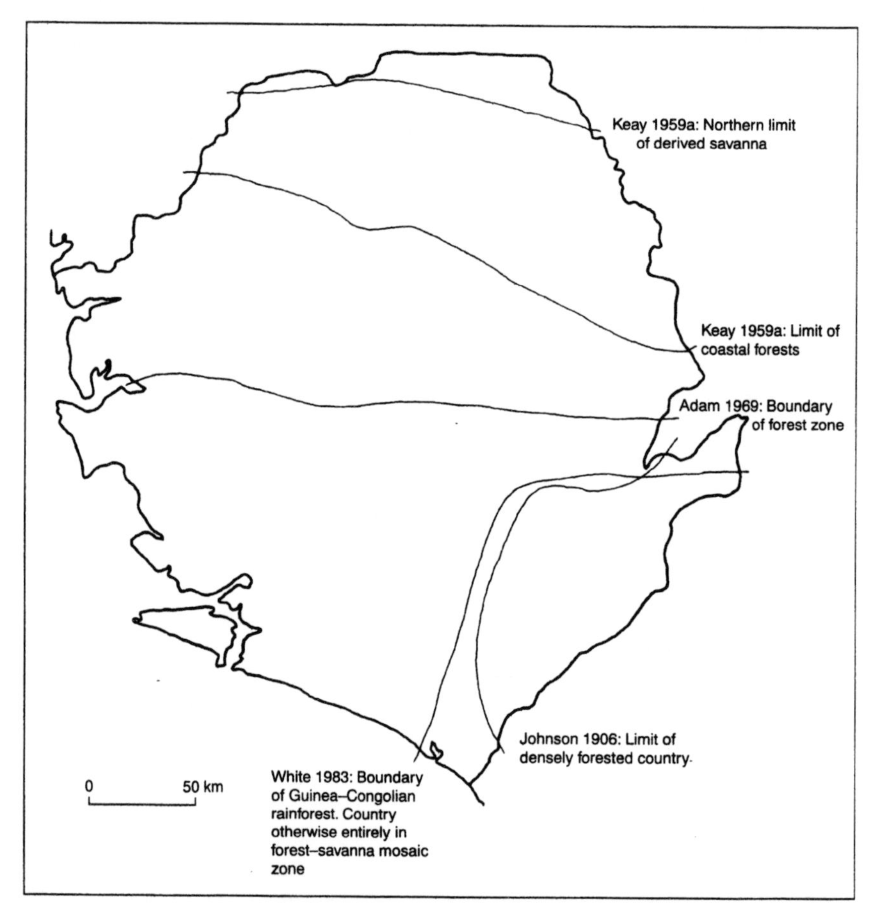

Map 7.2 Estimations of Sierra Leone's forest zone in early and modern works.

> [T]he Northern Province has always been spoken of as a region of savannah woodlands, open country, or orchard bush, but this is far from being exclusively the case. There are many forest remnants, areas which appear to exhibit the characteristics of true Rain Forest . . . on closer examination the forest of this region may not fall into the category of true Rain Forest . . . but . . . tropical semi-evergreen rain forest.
>
> (Government of Sierra Leone 1938: 9)

Although the direct path of influence is unclear, it is noteworthy that this re-analysis in Sierra Leone coincided with Aubréville's (1938, 1949) analysis of West Africa's past forest extent based on climatic reasoning, which also placed all but the extreme north-west of Sierra Leone within the forest zone. Keay (1959a) took up Aubréville's zonation in his influential map of West African vegetation, thus defining Sierra Leone's savannas as derived savanna. More recently, the UNESCO/AEFTAT map of African vegetation (White 1983) removed the historical deduction concerning the derived savanna zone, merely describing all of the country (except the extreme south-east) as forest–savanna mosaic. But strikingly the historical deduction has reappeared again in the recent presentation of Sierra Leone's vegetation by the World Conservation Monitoring Centre. This draws on White's zonation but states explicitly that his forest–savanna mosaic was forest 'at origin' (Sayer *et al.* 1992). At the extreme, Myers (1980) states that Sierra Leone's 71,712 km^2 lie almost entirely within the moist forest zone.

In sum, views which place Sierra Leone within the forest zone, thus giving it a forest past, have dominated in national and international policy circles and framed the analyses of many historians and social scientists. Yet the evidence presented from the timber trade, from vegetation indicators and from zonal classifications can, as we have shown, all be seriously questioned. Indeed, the counter-evidence presented here suggests the possibility of a very different vegetation history in interaction with land use; a possibility which strongly underlines the need for more site-specific historical research. It is in this light that we now go on briefly to examine a few sites in Sierra Leone where historical data have been collected and do question conventional ideas concerning gradual, unilineal loss of once pristine forest.

Site-specific cases of vegetation change

The Gola forest reserves

The Gola forest reserves which abut the Liberian frontier in the south-east of the country represent Sierra Leone's largest area of mature rainforest. As Kandeh and Richards (1996) argue, while the Gola forest commonly appears on conservation maps as primary rainforest, whether this should be so is a moot point. From the modern forestry and conservation perspective, several authors, notably the Sierra Leonean botanist Cole (1980), have considered most of the Gola forest reserves to be primary forest. Sayer *et al.* similarly assert that 'the moist evergreen forest in the Gola Reserve, which contains the last large remnant of lowland, closed canopy rain forest in Sierra Leone, is typical of climax Upper Guinean rain forest' (1980: 244).

Cole's analysis is supported by vegetation surveys whose findings were said to include several 'indicators' of primary status: (a) the massive size and height of emergent trees, especially the species including *Tieghemella heckelii*, *Heriteria utilis* and *Sacoglottis gabonensis*; (b) the high stocking rate of large trees; and (c) the occurrence of species dominance and species stratification, which is absent in young secondary forest. Indeed, in an overt way Cole uses present vegetation form as the basis for deducing vegetation history:

> The proportion of large diameter trees in a forest is a species characteristic as well as an indication of the past history of forest in much the same way that species stratification is seen in a forest only when the final heights of mature tree species are attained. The presence of species stratification and very-large-diameter emergent trees in the Gola Forest reserve (North) indicate forest maturity and a probable primary climax condition.
>
> (Cole 1980: 39)

Yet fifteen years earlier, when Small (1953) carried out surveys and oral histories he made a quite different interpretation of the same features which Cole takes to indicate primary status. Indeed, while it is difficult to be certain of locational comparability, both the vegetation species and forms picked out by Cole are singled out by Small as indicative of former farmland, leaving Cole's interpretation open to re-evaluation.

> In the period 20–50 years ago there existed many villages within the actual Reserve Boundary, the remnants of some of which are still visible. . . . The majority of the inhabitants of these villages were fishermen, but they had considerable rice and cassava farms within the forest. They are reported to have farmed the lowland areas and the marginal swamp areas, and, strangely, few farmed the steeper slopes. . . . The presence of large emergent species amidst the low, open canopied type of forest, prompted the question as to whether they would be remnants of old farms. This they affirmed to be true, stating that the species they left, i.e. *Sacoglottis gabonensis*, *Parinari excelsa*, *Parkia filicoidea*, *Amphimas pterocarpoides*, were too large or of no economic use. This, of course, would account for the large numbers and possibly distribution of very

> big *Sacoglottis gabonensis* within the area, with a notable absence of lower girth classes.
>
> (Small 1953, section 7)

Thus the species which Cole considered indicative of primary forest seem to be those characteristic of ex-farmbush, in which the canopy had perhaps become closed. Small also noted that large parts of the Gola forest reserves had been farmed during the nineteenth century. Small's impressions found some support in the 1923 report of the Government Lands and Forests Department:

> [T]he forest is virgin in character with the exception of large areas in the north-east and south-east. Here a close examination shows that the forest is secondary forest of considerable age, giving the appearance of virgin forest at first sight . . . strip surveys show a number of overold and over-mature standards that were left when farming was formerly carried on.
>
> (Government of Sierra Leone 1923: 42)

Migeod in 1926 came to the same conclusion, considering the Sierra Leonean part of the Gola forest to be

> mostly of recent growth, which growth became possible through the depopulation of the country in the last three centuries as a result of the slave trade which was in former times very flourishing here. There is every indication that the bulk of the forest here is of recent growth and expansion.
>
> (Migeod 1926: 331)

Earlier still, the forester Unwin made a tour of the Gola region in 1908 (Unwin 1909), noting the existence of large areas of forest in what are today the Gola reserves. The accounts he collected from inhabitants suggested that the area had been inhabited and farmed by Gola-speaking populations until around 1850. 'Many years ago, from all accounts, the Mendis attacked the Gola people, and drove them back across the Mano, so that now only the old foundations of the houses, and the Cola trees they planted, mark the sites of their towns, which must have been quite numerous' (Unwin 1909: 23). Oral accounts which Leach collected in Malema chiefdom in 1988 confirm these population dynamics and the presence of abandoned settlement sites in the north-eastern portion of the reserve. Further south-west, similar findings were reported by an interdisciplinary research team (Davies and Richards 1991; Kandeh and Richards 1996; Richards 1996) working from a base at Lalehun village, not far from the sites where Cole carried out his surveys and claimed to find primary forest.

In Gola it is not just the case that land which had been held in bush fallow rotations by previous populations has now reverted to its 'original' forest form. The legacy of past population and farming has influenced the character of the forest regrowth in an enduring way, not only in the features of tree species and size distribution noted by Small above, but also in the remnants of old cotton tree fortifications, of trees encouraged near settlements for their economic uses, and the legacy of highly fertile settlement soils with its potential enduring influence on forest quality. As Kandeh and Richards (1996) argue, patterns of biodiversity – for instance of rare bird species found in Gola – are closely bound up with the history of human settlement and need to be understood in relation to it.

Loma Mountains

The Loma mountain region provides a second area where historical evidence seems not to match vegetation histories deduced from interpretation of present vegetation. The area has been the subject of several botanical studies (e.g. Daveau 1963; Jaeger *et al.* 1966) which have generally identified an area of 'primary' forest on the mountain above 1,000 m, above secondary forest extending down to the plain; a distribution modified by the dry harmattan wind which reduces forest cover to the north and east. While they present the higher forest as untouched, the lower reaches are seen to be encroached by and under increasing pressure from gradually growing farming populations. Thus Daveau described Loma as:

> Avoided completely by man [sic] . . . for the most part covered by dense primary forest, above which emerge only the higher plateaux and summit ridges . . . of very special interest to botanists, who are able to find there relict vegetation types marking former stages in the floristic colonisation of West Africa. . . . The plain is the domain of wooded savanna, already Sudanian by virtue of its floristic composition. Its continuity is interrupted by temporary clearings made by the inhabitants of the rather isolated villages. . . . Above about 3,000 ft, on the other hand, the mountain appears to have completely escaped human interference. Appearances suggest that, here, there is primary forest, very rich in species, in which large trees are dominant and the underbrush is poor. The higher one goes, the more does one species, *Parinaria excelsum*, which grows to a height of between 60 and 80 ft, become dominant.
>
> (Daveau 1963: 2, 10)

Yet these accounts overlook some key aspects of population–vegetation dynamics which are revealed by historical data, both oral and documentary.

First, vegetation on the lower slopes and in the plain, inside and outside the reserves, appears to be ceding from savanna to forest. Oral accounts collected by Pocknell and Annalay (1995) from elders in the peri-reserve villages explained how the area of forest within their territories had increased during their lifetimes. Hills within the reserve now covered with young deciduous forest had previously been largely savanna woodland, as elders indicated by pointing out savanna species such as *Pterocarpus erinaceus* and *Daniella olivieri* within them. Second, elders identified the sites of old villages inside and outside the reserve. These signs of past habitation are evident within the areas treated by botanists as secondary forest, while they have not been looked for within those areas identified as primary. When farmers 'encroached' on reserved land for farming, this appeared to be less due to land shortage in unreserved village territory than to make use of these old settlement sites (Pocknell and Annalay 1995).

Third, the existence of abandoned settlements is unsurprising given that the Loma area has a documented history of warfare and depopulation. Laing (1825) during his 1822 visit to Falaba ascertained that by the 1690s, Yalunka and Kuranko people from the north-west of Loma were waging war with Kissi to their south-east, linked to slave trading. Kissi territory had once extended west, beyond the Niger sources, probably encompassing the Loma Mountains.[7] In about 1760 Kuranko launched a major invasion of western Kissi, and dry-season warfare continued so that by 1822, for example, heavy fighting around the Niger sources prevented Laing visiting them. By the colonial period, Kuranko had finally replaced Kissi in the Niger source region, leaving only a few isolated and politically subordinate Kissi villages there. Further depopulation seems to have occurred in the 1880s, linked to warfare waged by Samori Tourés' sofa warriors (see Fyle 1988a).

Thus while it is possible that areas of ancient forest have persisted on the massif, the history of this forest and of the lower slopes again seems not to be one of gradual encroachment under growing population pressure. Rather, the Loma area appears to have housed greater farming populations in earlier centuries, with increasing areas and maturation of forest following depopulation. As Pocknell and Annalay suggest, it cannot be assumed that this is mere forest 'recolonisation' of land which had been cleared of it, given Kuranko and Kissi land management practices which can encourage the establishment of woody fallows and forest in savanna (see Fairhead and Leach 1996a). It could be that following depopulation forest grew over land which had previously been savanna, encouraged by the legacy of such land use practices, possibly in conjunction with climatic changes. Such a case has been argued for what is now the Ziama forest reserve covering a different mountainous section of the Guinean 'dorsal' (Fairhead and Leach 1994a).

North-west Sierra Leone

In the region of the proposed Outamba–Kilimi National park in north-west Sierra Leone, deductions from present vegetation indicators have also driven views of vegetation history. Yet historical analysis has subsequently forced a revision both to the history and to the interpretation of its indicators, as demonstrated in the works of Nyerges.

The patches of semi-deciduous forest in this otherwise largely Guinea savanna landscape were, until recently, interpreted as evidence of a once more extensive forest cover, savannised by its inhabitants (Cole 1968; Nyerges 1988). In his early works, Nyerges (1988, 1989) presented a sophisticated analysis linking the ecological impact of swidden farming to the social organisation of fallow management. He argued that farmers in certain social positions fell back on forest-degrading practices, leading to irrevocable savannisation, and that the summary effect of such practices over several centuries had been large-scale regional deforestation. He argued (1987) that while population densities in Guinea savannna are low (typically less than 20 persons per km^2), ecological fragility is such that even under such populations the environment is already degrading slowly with traditional management. In essence, this analysis rested on observations of ecological and social processes over a single year, extrapolating their assumed effects backwards over decades and centuries.

Following more systematic consultation of early travel accounts, Nyerges has reconsidered this picture of progressive deforestation over the 80 to 100-year timescale (Nyerges 1996). In particular, Migeod's (1926) descriptions of the same area describe a landscape almost identical to today's. This is one area where the sources are particularly rich. Nyerges has begun to cross-check and put into time series descriptions such as Migeod's with other travel accounts such as O'Beirne to Timbo (1821), Laing to Falaba (Laing 1825), Watt and Winterbottom to Timbo (Watt 1794; Winterbottom 1803), Reade's two journeys to Falaba (1870, 1873), Blyden to Falaba (Blyden 1872), Zweifel and Moustier to Falaba and the Niger sources (1879), Trotter (1897), Scott Elliot (1893), and the accounts of Governor Cardew (in Alldridge 1910), to name but a few. In general these sources describe open grasslands, with forest patches either around villages or in regions recently depopulated. Thus Zweifel and Moustier, for instance, describe the land from Robanneh to Toncomba, and Yagalla to Nomoulaye as being 'desolate, nothing grows except some wild grasses, where far from each other one sees groups of trees' (Zweifel and Moustier 1879: 18). Blyden describes the route in a very similar way:

> The region of country through which we passed after we left Kambia, up to this point [Ganja], a distance of about one hundred miles, is of exceeding interest. It is mostly rich prairie land, with

> patches of heavy forest here and there. The soil in many places is of a rich, dark, fertile mould, supporting very luxuriant herbage and an ample supply of timber.
>
> (Blyden 1872: 6)

Slightly further north, during his expedition to Timbo, Blyden described how 'We set out from Turiyah and entered what is called the Tambaka country, and travelled for fifty miles to Sanyoyah through an almost waterless and uninhabited region. We found the land generally sterile, producing stunted trees and an inferior kind of grass' (1873: 122). Nyerges admits that 'the findings are obviously discouraging to any simple hypothesis of recent, large scale environmental decline and should serve as counters to any tendency to subscribe uncritically to "deforestation narratives"' (1996: 131). He reconsiders his earlier analysis of forest patches as relics of earlier more extensive forest, reinterpreting them as established around village sites, past and present.

However, Nyerges goes on to argue that early twentieth-century accounts 'imply nothing about older changes' (Nyerges 1996). He draws on the account of the Portuguese Jesuit priest Barreira of the nearby region of Bena to suggest that the area had numerous 'thick forests' in 1516, thus suggesting 'the reasonable interpretation that deforestation, if it has occurred at all, goes back some 500 to 700 years, rather than 100 or 50' (see Church 1968; Brooks 1993). However, as both Brooks and Nyerges signal, vegetation change and its causes at this timescale need to be examined in conjunction with climatic changes, in particular the dry period which Brooks identifies as occurring c. AD 1100–1500. Furthermore, Barreira's account needs to be interpreted with caution for several reasons. First, the Bena area may not be directly comparable with Nyerges' study site; certainly it carried forest early in the twentieth century when the latter did not. Second, it is unclear precisely what kind of vegetation Barreira was describing, or the exact referent of the Portuguese terms used. Third, it may not be valid to assume that the forests described were original, reflecting natural vegetation, especially given evidence that the area Barreira was passing through was an area of previously high populations decimated during the sixteenth century during warfare linked to what have been termed the Mane invasions (see Alvares 1615).

'Soil conservation areas'

Several authors, in particular Waldock *et al.* (1951) and Millington (1985), have attempted to isolate areas of particularly serious soil degradation in Sierra Leone; primarily where present land use observations suggest that grass fallows have replaced bush fallows. Here is not the place to investigate each of these, but one example will be illustrative. Of the ten priority soil conservation chiefdoms of Waldock *et al.*, four are in the eastern Kono border areas. Yet oral historical evidence from this region, supported by a spectrum of historical data from across the border in Guinea (Fairhead and Leach 1996a), suggests a history of vegetation enrichment in earlier grassland, linked to settlement:

> Our fathers then came and settled on the Melli River at Yawando, Yamba, Kongofie, and Sengi Sengi. No one lived in this country at the time, it was all grass, not bush at all. Elephants, lions, and wolves roamed everywhere, leopards also and all kinds of game. Our people gradually spread from the Melli river. . . . The rivers were then only small streams, and there was a scarcity of water. As our fathers had nothing to plant the farms with, they used to follow the tracks of the elephants, where after a time certain grasses used to grow in the elephants' dung, which they planted in the farms, and it was so that rice first came into this country. This is all so long ago that no man can count, but we, the old men, heard it spoken of by the fires in the evening when hugging our mothers' breasts.
>
> (Willans 1909: 141–2)

Such stories certainly reflect ecological components in political discourse, so cannot be taken at face value. But should they be ignored?

Dorward and Payne, in arguing for large-scale, agriculture-led deforestation in the twentieth century also single out two areas of particular soil and vegetation degradation as apparent evidence for this; the first is the Lophira savanna zone of the lower Rokel river; the second along the Kangari Hills. For the former, they assert that:

> No better example of the effects of deforestation can be seen than the present state of the country around the lower Rokel where the process first started in earnest. There in what was once high forest, are to be found vast areas of grassland and savanna woodland, the latter dominated by the small-growing *Lophira lanceolata*, a tree whose sole virtue is that it is resistant to fire.
>
> (Dorward and Payne 1975: 251)

No historical evidence is provided. Fortunately Laing passed through this region in 1822 and on his journey crossed 35 miles of 'extensive meadows, belted with thickets of wood about a hundred yards in breadth' (Laing 1825: 54). Nor do early accounts of the Kangari Hills support the conclusions which Dorward and Payne draw from the work of Nash (1948): that the area had recently lost extensive high forest cover. Lane-Poole in 1911 described the forest as 'confined to the upper slopes and top of the range' and stated that 'farmers have reached the upper slopes and have even crossed the range', 'that a few kola forests are preserved', but that 'exclusive of these they are at the mercy of the farmer, who is yearly driving his farms higher up the range' (Lane-Poole 1911: 8). Laing also passed by the Kangari and Sula Hills, describing them thus: 'The hills are clothed at their base with the camwood tree, and in places where it has been cut down, the sterile appearance produced by the multiplicity of stumps, is finely contrasted with the livid green of the smaller herbage and grass, which, interspersed here and there with a lovely palm tree, cover them even to the summits' (1825: 145–6).

Social and demographic analyses supporting views of deforestation

We are certainly not suggesting that no forest has been lost in Sierra Leone. But on the basis of cases such as those above, we do argue that in many instances assertions concerning forest loss are at the least questionable, and in others incorrect. The image of Sierra Leone as rapidly deforested, whether during the twentieth or nineteenth century, is deeply misleading. Furthermore, much high forest which does exist today has developed on lands once inhabited and farmed, then depopulated. In this context one is forced to question the ways in which demographic and ethnic arguments are forwarded to explain supposed deforestation, with Sierra Leone in this respect exemplifying arguments common throughout West Africa. Conventional understandings of population–land use dynamics in which forest recedes under steadily increasing population pressure are deeply flawed.

Sierra Leone's rapid population growth in recent decades, reaching high densities by West African standards, lends itself to Malthusian visions of population-induced forest loss and linked environmental degradation. Sayer *et al.* (1992) highlight that Sierra Leone had a density of 46 persons per km^2 in 1980 and 58 people per km^2 in 1992, with a 78 per cent population increase between 1963 and 1988, citing these figures to support their argument for rapid and recent deforestation and savannisation. But as Richards has argued, despite these densities most areas of Sierra Leone still have underutilised agricultural land, not least because government subsidies on imported rice and heavy marketing board taxes have provided disincentives to production (Richards 1996: 123). Even in the most densely populated rural areas such as around the Gola forest reserves, farmers' occasional reserve 'encroachment' has been motivated largely by politics – to restake claims over ancestral farmlands – rather than land scarcity elsewhere. Richards (1996: 118) also explodes the statistical 'myth' that fallow periods have suffered a catastrophic shortening in recent years, signalling a population–resource crisis. While it is often stated that fallow periods were 30–100 years in the past and have now shortened to an average of 7 years,

Richards points out that early sources in the region described even shorter fallows (e.g. 3–4 years, Dapper 1668; 4–7 years, Winterbottom 1803). Such assertions overlook established distinctions in sustainable local rice farming practice between long-fallow high forest farming and regular short-cycle bush fallowing.

Many analysts have also considered Sierra Leone's longer-term population dynamics in a unilineal way, as a settlement frontier moving into and then expanding within areas previously uninhabited or inhabited only insignificantly. Such analyses thus support – and are supported by – the image of large parts of the country as relatively pristine high forest 'at origin', gradually transformed by growing agricultural pressures. Colonial anthropologists and administrators commonly viewed the peopling of Sierra Leone in such terms. While certain groups, such as the Limba, were recognised as 'ancient' inhabitants of the country, many peoples were seen to have arrived no earlier than 1500, as part of waves of immigration linked to European trading presence and the break-up of Mande empires to the north. Sierra Leone was thus seen as part of the process summarised by Rodney: 'The peopling of the Upper Guinea Coast was a result of the continuous dislocation of population from the interior to the coast – a process that was largely precipitated by political events in the Sudanese states' (1970: 4). Population growth accelerated in the twentieth century under the assumed effects of 'pacification'.

The histories of particular 'ethnic' groups were written in terms which conformed to this view. Thus in his work in the 1940s, Little described the immigration of ancestors of the Mende into the forest zone in small bands from around 400 years ago, where they joined sparse populations of indigenous forest dwellers in a hunter–gatherer–trapping way of life (1967).[8] Hollins (1928), in an article on customary land law intended to inform the colonial administration, states that:

> The history of Mende chiefdoms cannot be traced back for more than a few generations. The origin of such chiefdoms into which the writer has probed starts – typically – from the clearing of bush and the founding of a town by some pioneer a hundred or two hundred years ago. Such a founder would seem to have come from the north with his followers – pushed south by the more warlike Mande people or north-east by the Golas and literally to have carved a holding for himself in the forest, driving out or absorbing the few autochthonous inhabitants and being joined by other clans or families, till there emerged what is now termed a chiefdom.
>
> (Hollins 1928: 25)

But while such Mende 'histories' complement conventional views of southern and eastern Sierra Leone as covered in high forest until a few centuries ago, then gradually denuded of it to leave only a few pristine relics such as the Gola reserves, they lie at odds with the alternative histories examined above; for instance, with a picture of Gola as intensely populated and farmed until nineteenth-century warfare.[9]

More recent work also invokes the image of a settlement frontier in Sierra Leone in recent centuries, progressively affecting areas previously occupied by forest-benign hunter–gatherers; for instance, in the historical geography of Siddle (e.g. Siddle 1968, 1969, 1970). But such views of recent population and agricultural transformation in Sierra Leone have been challenged by recent historical work based on early sources. Jones (1983a, b) argues that Sierra Leone has been relatively populous, with shifting cultivators growing rice, fonio and other crops on uplands for many centuries. Indeed, early sources speak far more of depopulation and of previously populous areas laid waste by disease and warfare than of unpopulated lands. For instance, in the Gola region there was probably earlier depopulation associated with the infamous Mane wars, as noted briefly by Manuel Alvares (1615). Depopulation also occurred as a result of disease. A detailed description of the Upper Guinean region from around 1630 writes of one such epidemic of dysentery:

> The Bloody-Flux begun in Sierre-Lions, in the year sixteen hundred twenty six, and spread itself through the whole countrey; raging with that violence and misery, that caus'd a dreadful mortality, and swept away such vaste multitudes, that for want of people, the rice tillage stood still above three years together; every one more dreading the day of his death, than making provision for the sustaining of such an uncertain life.
>
> (Dapper 1668, translated by Ogilby 1670)

Conclusions

Our above arguments concerning population–environment relations are certainly not meant to deny that periods of growth in farming populations have played a very significant part in transforming Sierra Leone's landscape. Nor do we deny that there are localities in which farmers face ecological problems because of the ways that population growth has combined with other production pressures. What is clear is that the unilineal views which have dominated so much of the literature obscure far more complex historical dynamics, in which populations have come and gone with varying effects on vegetation, including often overlooked processes of vegetation enrichment and shaping.

Sierra Leone's old-growth forest cover is, today, minimal. But far from having dwindled from 50 per cent or even 69 per cent since World War II, historical evidence clearly points to similarly minimal forest areas at the turn of the century. Nor, it can be argued, did the nineteenth century see the extent of timber-led deforestation which has been presumed. If the timber trade declined for lack of resources, it is probably because there were few there in the first place. How it is that images of such significant deforestation, often occurring within the lifetimes of present inhabitants, can be upheld despite strong counter-evidence is a question not only for Sierra Leone, but for all the countries discussed in this book; a question of no little practical import. It is to this question and to the power relations involved in imaging forest pasts that we now turn.

Notes

1 *Table 7.2* Recent estimates of forest cover in Sierra Leone

Date	*Area of forest*	*Source*	*Notes*
1976	626,200 ha, of which 365,200 ha closed canopy >30 m 261,000 ha 10–30 m	Gordon *et al.* 1979	Vegetation survey based on 1976 air photographs
1980	740,000 ha	FAO 1988	1976 air photo + model of intervening deforestation – includes mangroves
1987	400,600 ha	Sayer *et al.* 1992 (Paivinen and Witt 1988)	UNEP/GRID satellite data (does not differentiate primary and secondary forest and may have missed large areas of lower, more open secondary)
1990	756,000 ha	FAO 1993	denoted 'tropical rainforest'

2 Dorward and Payne then digress into a discussion of the distance of the camwood trade, which, as they state, did not contribute to deforestation. They use it to suggest ambiguously 'how far the range of the timber merchants extended'. It seems to us to suggest only how far it was profitable to carry camwood. Camwood, a special tree, is not used for timber and no evidence is supplied to show that its procurement was influenced by timber merchants, or that it was even traded by the timber merchants.

3 'Competition for this timber was so fierce that bitter feuding took place over single logs and small merchants were reduced to offering advance payment for uncut timber up to five, six or even seven miles from the river banks' (PRO CO 267/159. Gove Daherty to His Lordship John Russell, 22 April 1829).

4 While yields in some forests are considerably higher (e.g. 90 cu. m/ha in some Ghanaian forests, see van Rompaey 1993: 4), modern extraction ('mining') or

already depleted forests can produce much smaller yields of between 3 and 8 cu. m/ha (see Hasselmann 1986; van Rompaey 1993: 14).

5 This was the case for the great Limba towns of Konkoba, Katimbo and Kafugu on the Port Loko road, and Kamalafi, Bafudeyah and Kahko on the Kambia road.

6 We are grateful to Kate Longley for facilitating these conversations during her research in Kukuna, Sierra Leone.

7 On an 1840 Sierra Leone map, largely based on Laing, Kissi territory extends westward from Mount Loma, divided from linguistically related Limba territory by only 30 km of country marked as Kuranko.

8 Little's major work was published in 1967 but brought together analyses carried out largely during the 1940s.

9 This underlines the point that while oral accounts such as those heard by Hollins might well describe the political history of settlement and accurately convey the use of forest clearance as a foundation symbol, they cannot necessarily be treated as accurate representations of ecological or demographic history.

8

STATISTICS, POLICY AND POWER

For each country we have argued that the rate and extent of deforestation have been massively exaggerated by influential analysts from the early colonial period to the present. It is important to explore how such exaggerations became accepted and have, in many cases, persisted in scientific and policy circles. In this chapter, we examine how forest issues were analysed in imperial and colonial science, tracing the emergence of premises and methods in what was then the nascent science of forest ecology. These premises and methods, as outlined in Chapter 1, remain highly influential in modern analysis. We consider how forest conservation institutions and approaches were developed in accordance with this science and indeed were important to its production. We explore further how forestry analysis supported, and gained support from, colonial analysis of population, social and economic change. In short, we sketch the political and policy relations within which forestry knowledge was produced, and how alternative perspectives became obscured or marginalised in this process. Many of the broad relationships between power and knowledge apparent from early colonial times have, we want to argue, changed remarkably little to date, despite independence and developments in conservation strategy, ecological science and historical scholarship.

The emergence and institutionalisation of forestry science

The development of a pan-West African orthodoxy concerning deforestation can be traced at least as far back as the early colonial period, and the early attempts of scientists and administrators to comprehend vegetation status and change. Early colonial administrations elaborated maps describing the character of the vegetation in their countries. In various works between 1900 and 1933 the botanist Chevalier brought many of these together into a general description of West Africa's vegetation zones (1900, 1911, 1920, 1933b). This vegetation characterisation formed a basis for subsequent work by vegetation geographers such Schantz and Marbut (1923) and Mangin

(1924). Vegetation zones at this time were demarcated along several criteria: gradients which related spatial trends in species composition to trends in rainfall, relief and edaphic factors (Chevalier 1920; Aubréville 1938); variations in deciduous or evergreen habit and in presence or absence of grasses; and characteristic associations of particular species.

Chevalier's zones were originally based on descriptions of observed vegetation – or on patches of vegetation thought to be representative – but there nevertheless remained an ambiguity with zonal delineation based on the 'potential' vegetation which could exist under given climatic conditions. This ambiguity between actual and potential vegetation gave room for speculative deduction about vegetation history, by assuming that a zone used once to carry its potential vegetation. Early in the century, such speculative vegetation history was most strongly elaborated in the savanna areas on the northern margins of the forest zone. Early foresters and botanists considered that actual savannas were bio-climatically capable of supporting forest, and thus assumed that forest had once existed, having since been savannised through inhabitants' farming and fire-setting practices. This was the analysis of Unwin (1909) and Lane-Poole (1911) in Sierra Leone, Chevalier (1909, 1912b) in Côte d'Ivoire, Thompson (1910) and Chipp (1922, 1927) in Ghana, and Chevalier (1909, 1912b) in Benin. The work of Aubréville (1938, 1949), which correlated specific climates to specific forest types across Africa, helped to formalise these speculations, purporting to give a precise delimitation of this ex-forest zone within the savannas. It is this zone which became popularly known in anglophone circles as 'derived savanna', later appearing as such in the descriptions of Keay (1959a, b), Clayton (1961) and Hopkins (1965).

However delineated, vegetation zones provided a basis for defining 'natural' vegetation (and its local and regional variation), and hence for conceptualising subsequent modifications from it. The terms *'forêt primaire'* and *'forêt vierge'* were used in francophone circles, and 'primeval forest' in anglophone ones, to refer to what, in the wider discipline of ecology, soon came to be termed the 'climax vegetation'; the climatic climax being the maximum vegetation which a region's climate could support, and the equilibrium to which vegetation would return through succession following disturbance (Clements 1916).

The notions of balance or equilibrium in nature which framed these concepts had deep origins in traditions in western thought. George Perkins Marsh wrote in 1864 of the inexorable way in which nature, once disturbed, returns to balance (Marsh 1864). Yet by no means all ecologists even during the nineteenth and early twentieth centuries accepted these 'balance of nature' views. Charles Elton, in a textbook of 1930, argued that 'the balance of nature does not exist and perhaps never has existed' (Elton 1930: 16). Aubréville's vision of primary forest also did not accommodate equilibrium, but contained within it a dynamic turnover of species (Aubréville 1938).

Yet despite debate, it was on notions such as climax, equilibrium and succession that the emerging science of forest ecology came to be built.

The notion that there was a balanced, natural vegetation against which present, disturbed vegetation could be compared in turn supported the development of concepts and methods for defining and measuring departures from it. Scientists identified 'stages of degradation' – for example, those stages supposed to occur in the stepwise degradation of forest to savanna under repeated farming – so that the occurrence of any particular vegetation form came to indicate both a temporal degradation trend, and how far such degradation had got. Thus, by examining the species composition, diversity and associations (phyto-sociology) of a vegetation form, it became valid to deduce vegetation history, in ways which we have outlined in Chapter 1 and illustrated throughout the country chapters.

The possibility of basing vegetation history on deduction from present observations – of process from form – contrasted with the impossibility of using historical sources at the time. Not only were few such sources available, but their use was generally rejected. Aubréville, for example, was explicit, suggesting that the subjectivity of the term 'forest' and the indetermination of locality prevented comparison of present data with written sources (Aubréville 1938: 78). Oral history was generally used only to illustrate processes already known 'more scientifically', rather than being compiled in a systematic way. Opinion was more often solicited from white expatriates than from inhabitants.

Although these methods and deductive paths for determining vegetation history and the impact of land use were refined by early foresters for West Africa, they nevertheless depended on specific theories and assumptions which were current in European and Indian forestry circles prior to African colonisation. This is exemplified by the theories concerning the climatic impact of farming which were so endlessly re-elaborated in early West African forestry documents: that deforestation leads to climatic desiccation; that today's climate is in part drier than that historically due to deforestation to date, and that further deforestation will lead to further desiccation. Thus, when Chevalier envisaged a desiccation scenario for Côte d'Ivoire, he cited the pessimistic assessment of the French natural philosopher Poivre who had examined eighteenth-century deforestation in Mauritius (Chevalier 1909c; see Grove 1995). The centrality of climate to forestry and forest policy was clear in the work of Moloney, the first governor of Lagos (Nigeria), in his *Sketch of the Forestry of West Africa*, in which he analyses rainfall statistics for Sierra Leone, tabulating their steady decline over the four years 1878–82:[1]

> It is desirable that the attention of the community be drawn to the facts . . . showing a remarkable and steady decrease in the amount of rainfall in this district during the last four years. . . .

> The only cause that can be assigned for this decrease is the wholesale destruction of the woods and forests, which are at once the collectors and reservoirs of its water supply. This has occurred on other tropical regions, and when the cause was learned, by fatal experience through famine, the result of drought, then the forests were taken under Government protection and replanted, with the best results, but at great expense . . . I have added the rainfall statistics for 1883, 1884 and 1885 which point to an improvement in the direction of greater conservancy or more extended planting: perhaps of both.
>
> (Moloney 1887: 240–41)

The tight network of scientists within which such ideas were applied in West Africa helps explain the rapid establishment of a pan-West African orthodoxy concerning forest cover change. The network which contributed to coherence in British West African forest analysis was strongly linked to India and Burma. H.N. Thompson, the first trained forester in the Nigerian Forest Department, had earlier worked in Burma. Thompson was employed to detail Ghana's forests in 1908 and to found Ghana's department, whose first employee was M. McLeod, one of Thompson's subordinates in Nigeria and earlier trained in India. As Ghana's Forest Department grew, the trained foresters it hired – Chipp, Gent, Moor – had all served earlier in India or Burma. Indeed, Thompson specifically recommended recruitment from India. It was the second of Thompson's early subordinates in Nigeria, Unwin, also with experience in India, who was sent to detail Sierra Leone's forests in 1909, and to establish the Forestry Department. Unwin visited German Togoland in 1911 and by 1920 had written a major work on the forestry of West Africa.

Similarly tight networks characterised French West African forestry circles. Between 1900 and 1913, Chevalier made extensive tours of Guinea, Côte d'Ivoire, Togo and Benin. The uniformity in his reports became the basis for common policy in each of these countries, anyway linked within the colony of French West Africa. Chevalier also headed the Laboratoire d'Agronomie Coloniale from 1911, and the Mission Permanente des Cultures et Jardins d'Essais Coloniaux was soon established for him. His students authored early works on forest botany, ecology and history, including Fleury (concerning Cameroon and Côte d'Ivoire), Hedin (Cameroon), and Portéres (west Côte d'Ivoire) (Chevalier 1933b: 9).

Strong links were forged between French and British forestry services from the beginning, and were maintained throughout the colonial period. Thus, for example, Chevalier visited Thompson in Nigeria in 1905 and together they went to the botanical garden and Olokomeji forest reserve (Chevalier 1909c: 22). Thompson remarked five years later of 'my friend Monsieur A. Chevalier, who possesses an unrivalled knowledge of the tropical West African vegetation as found in its native haunts' (Thompson 1910: 69). Links were also exemplified in the Anglo-French Forestry Commissions of Nigeria/Cameroon and Nigeria/Niger from 1935–6 and in the Anglo-French Loma mountain visits in Sierra Leone (e.g. Adam 1949). Indeed, Anglo-French interaction was a main aim of the first Inter-African forestry conference in Abidjan in 1951, and of conferences of the West African Science Association (e.g. Alba 1956). French West African forestry and botany writing was known and quoted by British foresters, and British works by French foresters. In short, the world of forest policy and management in West Africa was small and analyses easily became standardised. As internationalism in forestry within West Africa grew, so standardised analyses were consolidated.

It would be wrong to suggest that the views of colonial scientists were homogeneous. What is clear, however, is that views which saw West Africa's forests and the impact of inhabitants' land use practices on them in less catastrophic terms very easily became marginalised. Thus, for example, many early foresters noted regions where forest was regrowing following depopulation (e.g. Unwin in Sierra Leone, Thompson in Ghana), and indeed many such areas were reserved. Yet with time and forest growth, the nature of past habitation was reinterpreted, with gradual marginalising of its significance to present vegetation. Equally, against many who asserted processes of unilineal savannisation on the forest margins, authors such as Vigne (1937) in Ghana, as we have seen, considered forest to be encroaching on savanna for climatic reasons. Yet his views were marginalised by Foggie (1953), who affirmed the same phenomenon but suggested that it was due to forest recolonising savannas following depopulation. Writings from the first decade of the twentieth century, even by influential botanists such as Chevalier, were notably less pessimistic than they later became. Indeed, some early colonial scientists were making observations which suggested inhabitants' enrichment of landscape; observations which we have drawn on in the country case studies. Yet the implications of these were ignored amid what became the dominant analysis of forest decline.

In other cases, individual observers who maintained ideas different from the mainstream found them difficult to pursue. By the time that analyses had been consolidated in the 1940s and 1950s, debates existed, but in a form which fine-tuned, rather than challenged, dominant theories; debates over the precise effects of bush fire, for instance (Jeffreys 1950), or over the soil effects of shifting cultivation. These were set within, and thus served ultimately to uphold, assumptions of rapid and unilineal forest loss.

The emergence of orthodox and standardised images of relentless deforestation were reinforced as they became institutionalised within West Africa's developing forestry administrations and policies. Throughout the colonial period, key forest scientists were also key administrators. The need for forestry departments was justified by these scientists largely on the basis of farmer-led deforestation and the threat it posed to climate and future forest productivity. In particular, colonial administrations saw a need to protect forests which could produce revenues, whether from timber or oil palms, and perceived an antagonism between local farming and these economic interests.

The form taken by forestry institutions, policies and legislation was shaped by the administrative contexts and possibilities of direct or indirect rule, but also reflected ways in which science and legislation had been merged elsewhere. In anglophone countries, Indian experience was influential. Thus in Ghana Thompson drew on southern Nigerian forest policy, which was itself based on the Indian Forest Acts (Thompson 1910: 114). French colonial forestry adopted aspects of the models used in France.

The very concepts central to early forest ecology were therefore concretised in forest policy and law, supporting – and coming to be supported by – the forms taken by forest reservation. Perhaps the clearest example is in the relationship between the notion of 'primary' (climax) forest and tenure. In proposing the reservation of the 'residual' forests of the Togo mountains, Aubréville explicitly linked the idea of primary forest with a justification for state possession, as under French colonial law the state had rights to 'vacant' land:

> Until recent years the inhabitants generally neglected these hillside forests. These are primary forests in which the local people have never exercised any right except for a few secondary uses such as the collection of fruit and other products. They are undoubtedly of the character of forests empty and without owners ['*vacantes et sans maître*']. The floristic composition enables one to say with precision whether a forest is virgin or of secondary origin. Only in the second case can the local people claim with any force to have even vaguely defined rights of occupation. Primary forest belongs to the State.
>
> (Aubréville 1937: 100)

Equally, early ideas concerning both ongoing savannisation and the relationship between forest and climate became embodied in the location of reserves and their management. The notion of ongoing southwards savannisation underlay the creation of 'curtains' of reserves in both francophone and anglophone countries, to defend the forests against savannisation. Such curtains were partially implemented in Guinea (Fairhead and Leach 1995b) and were established along the northern forest–savanna transition zone in Ghana, for example. Reserves were also created or planned around the headwaters of many rivers, to protect hydrological relations deemed important for maintaining 'normal' river flow. In Guinea a programme was elaborated in the early 1930s to protect the headwaters of the Niger river. In Ghana, several headwater reserves were established, protecting the Afram, Fum, Fure, Klemu, Ochi and Pompo rivers. Shelter-belt reserves were also established with the intention of humidifying farmland downwind. The relationship between forest and climate was inscribed at a more general level in the area of forest which planners aimed to reserve. It became a rule of thumb that in all parts of the forest zone 20 per cent of the area needed to be under forest (and reserved to ensure this) to maintain climatic conditions – a rule derived from eighteenth-century German forestry (Unwin 1920). This target was reached in Ghana in the colonial period, and is still striven for in Côte d'Ivoire.

In Ghana, tree tenure law in the 1949 forest policy also enscribed views of ecological history into legislation; it forged a strong distinction between natural and planted trees, with the state and its concessionaries asserting control over the former. In operation, this law defined all trees indigenous to the region as natural, and, where they occurred in fields or around villages, as relics of original vegetation. The policy thus served both to institutionalise the view that inhabitants were merely using, not actively managing, such trees, and to institutionalise the tendency to overlook landscape enrichment which might involve 'hitching a ride' on ecological processes (Richards 1987) without actually planting. Forest Department interest in controlling trees was mutually supportive of their scientific definition as relics of a former natural vegetation.

Forestry administrations were generally insufficiently strong until the 1920s for reserve establishment to proceed. Earlier local resistance had been highly effective, in anglophone countries, by preventing legislation (Grove 1994). In francophone countries forestry legislation enabled the state to take control over land but the nascent forestry administrations were incapable of enacting this. But by the late 1920s greater administrative capability, and at least in francophone countries military training and styles, underlay a major wave of reservation.

The financing mechanisms for colonial forestry departments, while varying nationally, served generally to support prevailing analyses of forest cover change and its causes. West African forestry services derived revenues from the sale of permits and licences for timber and wildlife exploitation, and from fines for breaking state laws. They were able to do this only by removing control over the management of resources such as trees, fire and areas of forested land from inhabitants, deeming the latter inadequate resource custodians whose activities, destructive of forest, required repressive regulation. Revenues were thus ensured by a reading of the landscape as deforested and endangered by the inhabitants who used it. In short, the economic and legal structures within which forest services operated can thus be seen to have helped frame the production of knowledge about forestry problems.

Convergences with analyses of social and demographic change

The analysis of forest cover change in terms of a steady and increasing decline also accorded with colonial views of other issues, most notably population history, the nature of African society, economy, ethnicity, and the character of migrants. In this sense analyses within the disciplines of botany and forestry were integrated with and mutually supportive of those within the emergent social sciences of the period.

As we have seen in the examples of various countries, linked to the idea of recent, intact forest cover was the presentation of the forest zone as only recently inhabited significantly by agriculturalists. The image was of a zone which once housed only sparse hunter–gatherer or minimal root crop cultivator populations with a benign impact on forest cover, awaiting the introduction of exotic cereal crops and iron technology as enabling conditions for population expansion, beginning gradually around 1500 and awaiting the twentieth century for their major impact. Slavery and internecine warfare were thought to have limited early population growth and, inversely, the colonial Pax Britannica and French 'pacification' to have unleashed it. Thus Sierra Leone's report to the first British Empire Forestry Conference stated that the 'primeval forest' is now 'almost disappearing. With the advent of the Pax Britannica . . . the native peoples have had a greater opportunity of multiplying and of devoting themselves to agriculture. Within a lifetime they have cleared up to about 90 per cent of the virgin forests by their system of shifting cultivation' (Thomas 1924: 5). In Ghana, for example, the colonial historian Ward stated that:

> There is no nation now dwelling in the Gold Coast which has been in the country much longer than the European. It may have been merely a coincidence that so many tribes were set wandering southward into the forest just at the time when the Portuguese seamen were clawing their laborious way from cape to cape round the West African shore. It hardly seems possible that the news of the Portuguese arrival can have helped to attract them into a land which had hitherto been almost, if not quite, uninhabited.
>
> (Ward 1958: 58; see Bourret 1960, Apter 1963)

Colonial anthropologists commonly contextualised the historical aspects of their work to accord with these prevailing images, as exemplified in the chapters covering Sierra Leone and Liberia. Population histories were commonly framed in terms of the histories of particular peoples, as constituted according to colonial perceptions of 'tribal boundaries'. Environmental behaviour was used as an important characterising feature in differentiating ethnic groups. In particular, Mande societies, whose origins were traced to the northern savannas, were identified as 'savanna peoples', in contradistinction to 'forest peoples'; long-term inhabitants whose origins could not be traced outside the forest zone and whose history was assumed to be forest-benign. These differentiations articulated with forestry discourse in at least two ways. First, the southwards migration of 'savanna peoples' was linked to the progressive southwards savannisation of the forest zone. Thus, in Guinea, Liberia and Sierra Leone the recession of the forest belt was seen to be accelerated by the migration from the north of Maninka 'savanna people' with a particular proclivity for fire-setting and forest clearance, as contrasted with 'autochthonous forest people' whose practices were more forest-benign. Colonial stereotypes concerning ethnicity could thus be linked to forest-related practices. As a result, analysis of forest loss could serve, in a mutually supportive way, to reinforce ethnicity, for instance, by feeding into local discourses concerning ethnic identity. Second, southwards immigration was seen not only to be expanding the population of the forest zone but also to be transforming its earlier, forest-benign economy by introducing new technologies (iron, cereals) and economic relations (trade).

Analyses which drew strong distinctions between 'original inhabitants' (autochthones) and 'migrants from the north' in turn became important in forest policy and administration. Forestry administrations sought to identify 'rightful owners' in negotiating reserve boundaries and, in the case of British indirect rule, administering reserves. The idea of autochthone or landowner was very useful in this respect. The construction of 'original inhabitants' implied, in turn, the construction of all subsequent arrivals as migrants, and different in their orientation towards forested land. Forestry policy commonly strove to limit the perceived nefarious impact of migrants' practices. In French West Africa administrators arguing for stronger forest law pushed the case that while in France even the rights of owners were restricted by state legislation, in Afrique-Occidentale française even immigrant, non-landholding strangers were unconstrained from abusing forest resources as they liked. As Sharpe has argued for Cameroon, forestry services may have helped in this way to concretise local social structures and the autochthone–migrant distinction (see Sharpe 1996).

In sum, by the end of the colonial period the analysis of forest cover change was embedded not only in contemporary science, but also in West African forestry laws, institutions and funding, and in the colonial representation of history. The colonial gaze on social and demographic history thus articulated with prevailing scientific theories and practices and their institutionalisation within West African forest policy, forming discursive structures which, we suggest, strongly conditioned descriptions concerning the causes, rate and extent of forest loss.

Post-independence continuity

The relationship between forestry knowledge and power can be seen in continuities between early and late colonial regimes, and those operating since independence. Where policy approaches have changed – as they evidently have in certain respects in the last few decades – it is instructive to examine the relationship between these changes and forestry sciences, and to consider how far they challenge earlier precepts.

Continuity in analysis and in policy approaches before and after independence owed a great deal to continuity in forestry administrations. In most countries, very little reorganisation of forestry departments took place at independence. In Ghana, Sierra Leone, Côte d'Ivoire and Guinea,[2] the senior officials who had held key colonial posts, if they did not remain in them, continued to be brought in regularly as advisory experts. They had trained other national officers for post-independence posts, which they subsequently took up. There was little change in the way foresters were trained, and forestry schools continued to use the same textbooks and training materials which had been developed by earlier colonial scientists.

In French West Africa the continuity was temporarily broken during the immediate period leading up to independence, when the repressive activities of forestry administrations sometimes became a focus of oppositional political struggle. Indeed, some forest reserves were de-reserved immediately prior to independence. But as new post-independence administrations gained strength and stability, so previous policy approaches were re-established if not reinforced, profiting, for example, from the new land nationalisation policies of independent governments.

This administrative continuity helps explain why there was continuity, too, in assessments of deforestation. In general, these continued to be based on methods which deduce history from current landscape form, excluding serious attention to historical data. By the 1960s and 1970s historical data were amassing and increasingly available, both in the form of longer time series for air photographs and new satellite images which could be compared with older photographs, and in possibilities to compare present vegetation with the documented accounts of early colonial scientists. Yet by the time such reasonable historical data had come to exist for West African vegetation history, the deductive paradigm had been set and the causes of deforestation ascertained. Subsequent generations of botanists and foresters had been trained, and the need to consult the historical record in examining vegetation change was not prioritised. There are several noteworthy exceptions to this: first, the highly influential study by Lanly (1969) in Côte d'Ivoire which suggested precipitous forest loss, but which, we have argued, was flawed in certain important respects that we have discussed earlier; second, a comparison made in western Nigeria by Morgan and Moss (1965) which forced a re-evaluation of assumptions concerning forest cover change, but whose implications were not widely considered. In general, however, historical data were deemed unimportant because the underlying assumption of recent, rapid one-way deforestation was so strong and so institutionalised as to render precision unnecessary.

Despite continuity in policy knowledge and assessment, historically inflected ecological research on forest–savanna transition issues in Côte d'Ivoire from the 1960s began seriously to reconsider the dynamics of the forest–savanna boundary. The need was signalled by Aubréville (1962), who in a dramatic U-turn reinterpreted his early work, arguing that the present forest–savanna boundary was in disequilibrium with prevailing climatic conditions which had become more humid. Forest thus had a tendency to encroach on savannas which had, in part, been climatically formed (albeit over a much longer timescale than our evidence would suggest). As detailed in Chapter 2, several authors identified processes of forest advance. Yet these have not been reflected in recent policy works on Côte d'Ivoire. The processes were either completely overlooked (e.g. by FAO 1981; Sayer *et al.* 1992) or dismissed (Monnier 1981) in view of other data concerning massive loss of forest cover. Even historians felt the need to reject oral evidence which they had collected concerning forest colonisation in recent centuries. As Blanc-Pamard and Peltre (1987) argued for Benin, the critique that Aubréville made of his highly influential earlier work has not, as yet, filtered through to the region's historians, foresters and conservationists.

As the country chapters document, works by historians and anthropologists which looked more critically at West Africa's socio-economic and demographic past were also amassing by the late 1970s and early 1980s. Yet these, too, have had little influence in conservation circles, where the views promulgated by earlier colonial observers have largely held sway. Historical analyses in modern conservation documents – for example, those in Sayer *et al.* (1992) – still make reference to colonial or immediately post-colonial historical and demographic analyses. This may be partly because the books and journals in which the more recent social science has been published are often not immediately accessible to those working in forestry and conservation. More importantly, colonial historical analysis was 'functional' to forestry institutions in a way that modern works are not. Many more recent attempts to describe deforestation and its social and demographic causes over the longer term amount not to critical analyses which employ data, but to illustrations of opinions already generally held (see Fairhead and Leach 1995a).

Continuity in administration and analysis has gone hand in hand with the persistence of certain specific policy and programme approaches. In many countries, colonial forest codes (which, as we have seen, incorporated particular analyses of forestry problems) have been retained until very recently. Ghanaian forestry, for example, was governed by the 1949 Forestry Act right up until 1996. Inherited forestry codes brought with them inherited assets, whether in forest reserves or listed timber trees. In their continued presence, these assets, in effect, constantly instantiate the original reasons for their acquisition. Thus, the existence of reserve curtains reinvokes the idea that southwards savannisation is in progress. Watershed reserves continually reinvoke the supposed links between deforestation and hydrological desiccation. The presence of state-controlled timber trees justified as such by their supposed 'naturalness' continually reinvokes the idea that such trees in farms and near villages are indeed natural relics of a lost forest cover. For forestry administrations to relinquish the analysis linked to these assets would be, in effect, for them to relinquish claims to the resources. The analysis is thus strongly implicated in the real politics of control over valuable resources and revenues. As Roe has argued, crisis narratives – in this case about deforestation – are a 'primary means whereby development experts and the institutions for which they work claim rights to stewardship over land and resources they do not own' (1995: 1066).

There are many examples of continuity even in conservation programme ideas, whether through inheritance or because administrations have come up, anew, with the same idea. In French West Africa, a major programme to reserve and reforest the upper catchments of the Niger river in order to regulate its downstream flow was first mooted in 1932. Apparently forgotten, it was raised again in 1950 but again left unfunded. Apparently forgotten once more, it was finally raised again and funded by international donors in 1990. 'Sacred groves' have also been a recurrent theme in conservationist thinking. As early as the 1930s senior forestry administrators were commenting on the ways that local religious and cultural institutions apparently served to preserve 'islands of nature', linked variously to rituals, initiations or ancestral activities, in an otherwise devastated and desacralised landscape. Even at this time they speculated on ways that such sacred groves could be harnessed as part of conservation strategies. Since the mid-1980s, it seems, the sacred grove theme has risen to greater prominence especially among international donors, who are now supporting a host of government attempts to intervene in and support their management to enhance forest and biodiversity conservation, including attempts to recognise them within state law. These recent projects, like their colonial predecessors, are grounded in and perpetuate the assumption that the forest patches in question represent an original, natural and endangered vegetation.

In some instances, administrative changes in the post-independence period have served to strengthen accepted analyses of deforestation as recent and one-way. As the power vested in the state grew and consolidated following independence, governments were sometimes able to acquire greater control over forest reserves than had originally been taken from inhabitants, sometimes by designating areas as strict nature reserves or national parks. Inflation reduced local benefits from and economic stakes in forest reserves still further, by reducing the financial rewards that chiefs and communities might have gained from logging royalties to a pittance. Meanwhile over time, the forest within reserves grew and developed further, so that even those which had been 'new' forests – of recent secondary growth following abandonment from farming – at the time of their original colonial reservation came more and more to resemble the image of pristine, climax forest. As the impact of previous populations became less visible, so their history became more opaque and easily 'forgotten', if even recognised, by reserve administrations. As forest grows, so the idea that people have inherited rights grows more distant.

These various forms of inertia and consolidation, we would argue, help to explain why 'new' ecological thinking has apparently made so little impact on forest and conservation policy. For in the academic field of ecological science, understandings of forest ecology have undergone some fundamental changes in the last few decades. It was in the late 1950s and early 1960s, for instance, that new views began to take shape concerning the relationship between forests and climate. In contrast with previous notions of climatic climax, these views suggested that forest distribution, form and species composition might be responding to long-term climatic change and therefore out of equilibrium with present climatic conditions. Aubréville, for instance, completely changed his ecological analysis between 1950 and 1962 in this respect; yet none of his later analysis made any impact on policy. More recently, there has been a consolidation of scattered earlier work which replaces the idea of a stable, balanced 'nature' with ideas of constant variability over time and space, and of non-equilibrial forest dynamics (e.g. Botkin 1990; Sprugel 1991). Such approaches – in line with what some have termed the 'new ecology' – have fundamentally challenged earlier theories of climax and succession, and of the dynamics of disturbance and recovery. Notably, these newer perspectives in ecological science are grounded in an historical understanding of ecology, and have developed alongside greater attention to historical sources. They are not generally based on a reformed understanding of people's role in forest dynamics, or any greater appreciation of indigenous vegetation management practices. Nevertheless, by emphasising the extent to which forest presence and composition is the outcome of historical interactions between multiple factors and processes including major disturbance events, or even the result of unique pathways of vegetation change, they could allow also for radically different interpretations of human–vegetation interactions (Fairhead and Leach 1996a). Yet their impact on policy has been minimal. It appears that a dislocation has emerged between the science of forest ecology, and the analysis which continues to inform conservation policy.

Where new ecological ideas have come face to face with conservation policy, the uneasy relationship between them is clear. McNeely (1994), for instance, incorporated recent thinking about forest ecology and history into a questioning piece which, in asking 'what is "natural" vegetation in a changing environment?', explicitly removed many of the conventional justifications for forest conservation policy. In pointing out that forest landscapes are a continually changing product of interactions between people and ecology, the moral and scientific grounds for intervening to restore a more 'natural' vegetation are replaced by difficult social and political choices about whose vegetational priorities – including inhabitants' own – are to prevail. An UNRISD Discussion Paper (Pimbert and Pretty 1995) which made the case for far more active participation by and tenurial rights to forest-edge populations on the grounds of their historical roles in shaping forest vegetation was very badly received by major conservation agencies, partly, we must assume, because of the threat it posed to established conservation approaches. In the supposed state-of-the-art review of African forestry issues put out by major forestry conservation donors and policy organisations, 'new' ecological thinking is conspicuous by its absence. The IUCN atlas of African forests (Sayer *et al.* 1992), for instance, continues to use a language of 'original' and 'climax' vegetation, failing to take on board more recent concepts or research on people–forest dynamics and climate. Where it does address climate change, it sets this so far back in time (c. 8000 BP) as to be largely irrelevant to present forestry concerns, while overlooking the major evidence now available for more recent and significant climate change.

Shifts in forestry and conservation policy agendas

Room for new analysis?

It is instructive to examine a number of arenas where there have been shifts in forest and conservation policy and priorities over the last few decades, and to explore their links with modern scientific enquiry. In many cases, supposed new approaches and areas of concern in policy have been accompanied by new foci and emphases in science. But they have not, we argue, brought about shifts in the basic conceptual apparatus and assumptions underlying deforestation analysis, or led to greater attention either to inhabitants' knowledge or to new ecological thinking in policy circles. More often, older scientific ideas have simply been reinforced and put to work in new contexts.

As a concession to greater 'participation' from local populations in forest conservation, many agencies have, since the early 1980s, promoted the approaches which have come to be known as 'conservation with development'. Indeed, variations on the Integrated Conservation with Development Project (ICDP) model now dominate donor support to West African forest reserves, including for biodiversity conservation. In strong contrast with the exclusionary approaches dominant in the colonial period, these are based on the view that local populations should co-operate with and participate in reserve management. While equity concerns are part of the justification for such approaches, they are evidently also underlain by questions of feasibility and efficiency. Experience during and since colonial times has shown that where exclusionary approaches generate local resistance, poorly staffed and financed forestry administrations are often simply unable to control 'encroachments' on reserves. Instead, the approach is now to ensure local co-operation and support for reserves, generally through allowing forest-edge inhabitants some material benefit from them, whether in the form of rights to collect forest produce or ecotourism revenues, or in 'compensatory' rural development in reserve buffer zones (Wells *et al.* 1992; IUCN 1994).

Conservation-with-development approaches certainly offer the potential for greater attention to inhabitants' concerns, and for supporting rather than inhibiting local resource control as did earlier exclusionary approaches. That, in practice, many such projects have failed to secure local support relates – as a growing critical literature now explores – partly to issues of implementation, partly to the tokenistic nature of 'participation' sought, and partly – and of relevance here – to flaws in their underlying conceptual apparatus (e.g. Gomez-Pompa and Kaus 1992; Pimbert and Pretty 1995; Richards 1996). Conservation-with-development approaches tend to draw on, and thus reinforce in a new policy context, particularly ahistorical views of society–nature relationships. They generally exhibit no revision of earlier views that forests and their valuable biodiversity represent 'nature' of a more or less pristine sort. The issue, as before, is how to make people's impact on natural forest more benign, not how to work with it to make 'nature' more diverse and socially useful (see Guyer and Richards 1996). Moreover, the approaches reveal an ahistorical conception of social form which sees people's need to use forest produce and desire to farm land in reserves as phenomena driven by population pressure, recent social change and economic need. In many senses, they merely extend in a modern policy context the tenets concerning recent frontier-like pressure from growing, immigrating, modernising populations which, as illustrated in the country cases, have long dominated social science arguments concerning deforestation. The solutions suggested, and narratives which justify them, exclude from consideration other aspects of landscape history: for instance, that reserves might be ex-farmland in which encroachments represent attempts to reclaim ancestral lands, or that forest product collection sites are valued for their significance as products of earlier settlement, as well as for their present resources. By failing to take on inhabitants' own debates through such exclusions, the approaches may compromise their own success in 'participation' as well as serving to entrench outdated scientific perspectives.

Similar arguments apply to recent policy attempts to promote 'participatory' management of smaller forest patches as 'community forests'.

Fuelled by 1990s concerns with decentralisation, democratisation and the view that sustainable development should be based on local initiatives (e.g. IUCN/WWF/UNEP 1991; Holmberg *et al.* 1993), many donors and governments have promoted forms of joint or collaborative forest management in which government recognition and support is given to local control and management of forest areas. In West Africa, the Ghanaian Collaborative Forest Management Unit, the village forest programmes in Guinea and the inclusion of forest patches within francophone *gestion des terroirs villageoises* (GTV) programmes exemplify variants of such approaches. While differing in strategy and administrative/legal context, however, these approaches tend to subscribe to a common underlying narrative. In this, 'the community' (commonly viewed as a rather homogeneous whole) is assumed, at some time, to have existed in balanced harmony with an undisturbed forest environment; a harmony disrupted in more recent times by the combined forces of population growth, commercialisation, migration, social change and modernity. The task of community forest management is to bring people and environment back into harmony through re-establishing or rebuilding 'community' institutions. In effect, the approaches reinvoke much longer-established images of African communalism in harmony with nature in the absence of economic and social change. They also make use of colonial anthropological views of functional social organisation, again reinforcing these in a new context. Arguably, these overly static conceptions of people and environment have served to compromise the success of such approaches (Leach *et al.* 1996).

A contrasting donor and government focus with a more individualistic rather than communal emphasis has been on farm forestry and agroforestry (e.g. World Bank 1991). Driven perhaps less by forestry professions and disciplinary perspectives than by agriculture and agronomy, agroforestry research has nevertheless led to new foci in this rapprochement; to a science of farm trees, tree–crop, tree–microclimate and plant–soil interactions. It has been accompanied by a socio-economic literature concerned with production pressures and farmer incentives; and of tenure and labour at the farm level (e.g. Bruce 1990; Prinsley 1990). The focus on small farmers has encouraged this area of policy and research to pay relatively more attention to indigenous knowledge. Nevertheless, the policy debates tend to draw on and reinforce distinctions between natural and planted trees. Attention is on indigenous knowledge as a science of planting, and of valuation of existing trees; a planted–wild distinction which obscures attention to the ways that farmers work with ecological processes to encourage trees to grow; to influence them whether by settlement, changing soils, altering fire, etc. Equally, socio-economic arguments link farmers' control, tenure and incentives to planted trees, as distinct from naturally occurring ones. By focusing on individual intentionality, the debates exclude attention to ways that landscapes might be enriched with trees through the cumulative effect of social processes such as settlement, and the more complex layers of tenure and social institutional control linked to such landscape history.

Most fundamentally, the drive for farm forestry and agroforestry research and programmes, as well as community woodlots, has generally emerged in the context of concerns about forest loss, and thus serves to reinforce the views underlying these concerns. Whether governmental or non-governmental, they are often justified and gain their funding in these terms. Smallholder tree-planting programmes commonly aim to alleviate pressure around forest reserves; to reforest denuded watersheds to reinstate downstream conditions; to re-create or maintain lost or threatened soil fertility, or to reforest degraded reserves. Whatever the context and whatever the particular approach, the common tendency is for programmes to be constructed (and their funding justified) against a backdrop of decline; to mitigate with 'improved' land use practices, systems, economic incentives or social organisation, forest degradation which is seen to have taken place.

The changes in approach not only reinforce ideas of decline, but also accompany an extension of state control over forest resources. They extend forest department interests and powers into lands which are being newly reserved, albeit under different (and 'participatory') management regimes, and out of reserves into buffer zones and on-farm land management. 'New conservation' may have participatory intent, but often has the effect of further resource alienation and the expansion of state bureaucratic control – albeit nowadays with NGO assistance (see Ferguson 1990).

Conclusions

In this chapter we have sketched some central links between ecological representation and political/institutional form; links hinging on control over material forest resources, and on the allocation of responsibility to decide their management. In this respect it is a sketch of 'successful' policy knowledge; in other words, knowledge which was incorporated into, shaped, and was in turn shaped by forest policy. Our account is therefore partial; it does not attempt a full 'history' of knowledge about forestry in West Africa, something which would demand greater attention to differing interpretations, locating these in the politics of the time. Such a history would also require more attention to ideas concerning forests among non-foresters, in particular those working in agricultural and other administrative departments, and to the assorted ways in which inhabitants' perceptions articulated with those of scientists and policy-makers. Equally, it would need to accommodate the representations of forests in wider literary and popular culture. This fuller history would necessitate a level of detail which can generally be captured only in case studies, such as we have attempted in earlier works (Fairhead and Leach 1996a).

Our partial representation cannot therefore be taken to suggest that there has been a single hegemonic vision concerning forest issues in West Africa. Nevertheless, the extent of exaggeration of deforestation across countries, and its continual rescheduling to the present, has required us to acknowledge and explore the political context in which this knowledge has been produced and shaped. In the conclusion, we ask what alternative approaches to research and policy might be suggested by the analysis presented in this book.

Notes

1 In much of this he is quoting a Dr Hart of the *West African Reporter*.
2 Liberia, of course, had no such administrative disjuncture, at least until 1980.

9

REFRAMING FOREST HISTORY, REFRAMING FOREST POLICY

Each of the country cases considered in Chapters 2 to 7 has highlighted different issues in the study of vegetation change. The case of Côte d'Ivoire underscores how the misuse of statistics can produce catastrophic, but erroneous, assessments of forest loss. Moreover, foresters (and even socio-economists) have premised their analyses on an assumption of minimal Ivorian populations in the nineteenth-century forest zone, overlooking far more complex population–resource dynamics and important instances of forest expansion in certain times and places. The Liberia case emphasises how a unilineal vision of forest conversion at a 'frontier' can take hold in academic consciousness across a range of disciplines, yet is probably a fundamentally misleading model for understanding vegetation dynamics, and obscures important aspects of Liberian history. In Ghana, influential international analyses have produced an image of the entire bio-climatic forest zone as having been covered with 'original' forest at the turn of the twentieth century. Yet there is strong evidence that much of today's forest is of relatively recent, and perhaps even anthropogenic, origin. In Benin, scientists and policy-makers have interpreted forest islands, isolated forest trees, baobabs in secondary forest thicket, and the presence of extensive palm groves to indicate degraded tropical forest. Yet the weight of evidence suggests that inhabitants past and present may have established each of these vegetation formations in open savanna country. Easy deductions cannot therefore be made. The Sierra Leone case exemplifies how a faulty analysis of technology and socio-economic change, framed by and reinforcing views of deforestation, can spread and become magnified within academic consciousness. Whereas the supposed role of the timber trade has dominated analysis in Sierra Leone, in Togo other technologies – iron smelting and intensive agriculture – have been blamed for deforestation. Yet re-evaluation of the evidence provided suggests that historical iron smelting might have been more sustainable than recent analysts have considered it to have been. Intensive agriculture involving forest clearance in certain areas is shown to be responsible for forest establishment elsewhere.

It has been instructive, we think, to examine these cases together. For while debates have developed in rather different ways in each country, reflecting their particular administrative and intellectual histories, as well as the substantive ecological and socio-economic issues on which the analytical and policy gaze have fallen, there are clearly obvious similarities in the ways that policy knowledge has been elaborated. Recurring themes are revealed more clearly when country orthodoxies are interrogated together. Dwelling on each country alone would limit this potential for re-evaluation. As we sum up in this conclusion, the group of country cases adds up to more than the sum of their parts, and highlights the need for new research and policy perspectives on West Africa's forests.

The extent of deforestation

Deforestation during the twentieth century has been significantly exaggerated in every country covered by our research. Each chapter has considered how the production of deforestation figures is riven with definitional problems, but has nevertheless shown that figures in circulation are influential both nationally and internationally, yet are wrong.

Table 9.1 compares current orthodox estimates of the extent of conversion of dense humid and semi-deciduous forest to farmbush and savanna during the twentieth century, with our revised estimates as suggested by each country study. It suggests that twentieth-century deforestation in these countries is probably only one-third of that suggested by the estimates in international circulation. If our assessments are even approximately correct,

Table 9.1 Suggested revisions to deforestation estimates since 1900 (millions of hectares)

Country	*Orthodox estimate of forest area lost*	*Forest area lost according to World Conservation Monitoring Centre**	*Suggested revision*
Côte d'Ivoire	13	20.2	4.3–5.3
Liberia	4–4.5	5.5	1.3
Ghana	7	12.9	3.9
Benin	0.7	1.6	0
Togo	0	1.7	0
Sierra Leone	0.8–5	6.7	c. 0
Total	25.5–30.2	48.6	9.5–10.5

Note

* The World Conservation Monitoring Centre estimates (Sayer *et al.* 1992) are of forest loss from an undated 'origin' forest cover and are extraordinarily high compared with any other available source. Hence they are presented separately from other orthodox estimates.

the exaggeration of early forest cover and of subsequent deforestation must be significant to those who model the region's climate, greenhouse gas emissions, biodiversity and ecologically influenced epidemiology. As incorrect panel data, they will mislead, in turn, cross-country analyses identifying deforestation causes, and hence the modelling of forest futures. At the same time, they provide misleading guidelines for more locally grounded forest and land use policies.

Population, depopulation and new forests

It also seems likely from the cases that 1900 to 1920 – the baseline nowadays commonly taken for deforestation assessments – was in itself a high point in forest cover for many West African countries. We have forwarded evidence to support this argument for Liberia and Ghana, and have suggested that it may also hold for Côte d'Ivoire. Much of the forest cover present in 1900 had grown relatively recently over lands inhabited and farmed during earlier centuries. As a result, much of the forest loss which did occur during the twentieth century was of such 'new forest'. Estimates of forest loss for these countries would, therefore, be significantly lower were it possible to take an earlier baseline.

Many of the forest areas which were reserved in the colonial period, and which are today valued as key biodiversity and timber havens, also have a history as new forest. Recognising such dimensions of vegetation history underscores the point that it is wrong to approach West African forests as if they were 'primary' or 'virgin', coming under pressure for the first time; an image frequently encountered in modern policy literature. Even those foresters and policy-makers who have accepted that the forests they observed were 'secondary' have tended to minimise this history as ancient and forgotten. Yet as we have argued, depopulation may in fact be relatively recent, and hence of real importance for understanding not only forest ecology, but also inhabitants' attitudes towards forest. Processes leading to depopulation – occurring only a few generations previously, in many cases – may rest strongly in the social memory and cultural practices of a region's inhabitants, inflecting their attitude towards the forest. For example, forest policy documents repeatedly lament farmers' penchant for clearing high forest even when farmbush fallow is available, whether attributing this to the greater yields they obtain or to the affirmation of masculinity in felling high forest trees. But an alternative explanation would be that in clearing high forest, farmers are attempting to maintain or restake claims to ancestral lands. More broadly, the discursive denial of history apparent within state forest and agricultural policies obscures processes in political history and cultural memory which may be central to understanding the recent civil unrest in several countries of the region (see Richards 1996).

Understanding the ecology of forests and forest–savanna transitions

Exaggerated claims concerning the nature and extent of past forest cover and of subsequent deforestation have, we would argue, been misleading ecologists. Present forest structure, species composition and ecological interactions may reflect less 'nature and its degradation', than real histories of climatic fluctuation in interaction with earlier vegetation forms and land management. In particular, in West Africa, claims of one-way deforestation have completely obscured what seems to have been a large increase in the area of humid forest formations and secondary forest thicket over savanna in recent centuries; a finding for which we have shown there to be strong evidence in Sierra Leone, Côte d'Ivoire and Ghana as well as Guinea (Fairhead and Leach 1994a, 1996a). Indeed, this process appears to be ongoing (Lambin pers. comm.). In this respect, evidence from vegetation history supports the arguments of climate historians such as Nicholson concerning recent rehumidification in West Africa's forest region, trends which, as we suggested in the Introduction, may be the inverse of those in more northerly, drier zones.

An historical approach, then, forces new questions to be asked concerning the status of both the forest–savanna transition zone and the semi-deciduous forest zone to its south. As we have argued elsewhere, there are strong grounds for rethinking the forest–savanna mosaic – usually seen as the result of forest loss and savannisation – as the outcome of forest expansion under the complementary forces of human land use and climatic rehumidification (Fairhead and Leach 1996b). As the forester Hawthorne suggests on the basis of species diversity, it may be more appropriate to consider the semi-deciduous forest zone 'not as an intricately balanced ecosystem likely to fall apart after minor disruptions but more nearly an ad hoc assemblage of species thriving after millennia of disturbance' (Hawthorne 1996). As he commented on an earlier draft of Chapter 7 (Ghana), the historical evidence presented here 'adds flesh to the bones of a theory that the entire semi-deciduous forest zone is anthropogenic in a fundamental sense' (pers. comm.). Tree species characteristic of the drier semi-deciduous forest belt, such as *Milicia excelsa*, *Antiaris toxicaria*, *Triplochiton scleroxylon* and *Celtis* spp., which are usually considered to be characteristic of 'secondary forest' and of forest in the drier regions, might better be interpreted as characteristic of zones of recent forest expansion.

The work of Hawthorne exemplifies how our attention to historical sources in this study dovetails with an emergent trend in the science of forest ecology. A number of scientists – to date largely examining temperate regions – now work with the theories in 'new ecology' raised in Chapter 8 to consider forest composition not in terms of divergence from a 'natural' equilibrium, but as the outcome of continuous transformations influenced by multiple interacting factors, disturbance events and legacies. Consequently, forest composition at any moment has to be considered in terms of the unique pathways which led to it, in particular spatial and temporal contexts (e.g. Sprugel 1991; Peterken 1996). At a general level, for instance, there is a tradition of examining West African forest composition in terms of climatic forest gradients (i.e. the pattern of distribution of different forest tree species in relation to climate).[1] But those working from new ecological perspectives ask further whether these are gradients only in space or also in time, reflecting climatic changes and other influential events.

Many ecologists consider this type of rethinking to require attention to history and disturbance events in climatic and other ecological processes. But they continue to ignore the multi-faceted influences and legacies of people's land use and management. The incorrect social history on which forest ecologists so often rely has encouraged this tendency. Given that forest ecologists are using incorrect analysis of historical demography in Ghana, Sierra Leone and Liberia, for instance, they are better able to ignore human history in understanding modern ecology. Interpretive paths are closed off by views of historical demography which present the forest zone as disturbed only recently. The evidence assembled in this book exposes the problems in circumscribing analytical concerns in this way. It shows how an appreciation of the impact of past populations, settlement and land use suggests factors crucial in interpreting present-day vegetation forms.

Anthropogenic influences on vegetation

Each chapter has exemplified in different ways the anthropogenic shaping of West Africa's forest vegetation, whether as the unintended consequence and legacy of everyday life, or in purposive management – where people have actively altered vegetation to their own ends.

The enduring effect of settlement on vegetation is fundamental. Inhabitants in their everyday lives can render soils highly fertile, and manage village peripheries in ways which shape tree cover. Historically fortification has been important, but land management for kitchen gardening, multistorey home gardening, fire protection and ritual purposes has also served to create patches of forest in savanna, and patches of characteristic forest within existing forest and fallow formations. The effects endure long after a settlement site is abandoned. While evidence of such processes can be found in savannas and forests throughout the region, it has consistently been ignored in analyses which have used these rich and impressive vegetation patches to represent the region's natural vegetation.

Farmers frequently encourage particular tree species to grow in their fields or settlements, preserving, planting or transplanting them whether for their economic values, beneficial effects on soils and crops or ritual purposes. Vegetation composition is frequently the outcome of such active management of particular trees. We have noted many instances where isolated trees of forest species are more plausibly interpreted as volunteers or enriched products in farmers' fields or old fields, than as relics of earlier vegetation.

Equally generic to the region is the creation of more tree-rich fallows through certain forms of farming. While it has been assumed that shifting cultivation has been responsible only for the conversion of high forest to fallow, and for progressive deterioration and savannisation of fallow, the scattered instances which we have assembled indicate that the reverse can be the case: farmers can and have upgraded savanna to forest thicket fallows, or to palm forests and multi-storey home gardens. Cases such as those presented from Togo (Guelly *et al.* 1993), Côte d'Ivoire (Spichiger and Blanc-Pamard 1973), Ghana (Amanor 1994a) and Sierra Leone (Pocknell and Annalay 1995) and by Fairhead and Leach (1996a) for Guinea may, when considered on their own, seem like the isolated cases that their authors assumed them to be. But together they demand to be incorporated into a more fundamental revision of the interpretation of West African vegetation forms. Further research on farm–fallow interactions from such a refocused perspective would, we suggest, be likely to reveal many more similar cases.

When the effects of climatic change are considered over such shorter timescales, then their interaction with anthropogenic factors becomes more relevant (see Fairhead and Leach 1996b). Indeed, we have suggested that particularly at the northern margins of the forest zone, people's land use practices and climatic rehumidification may be complementary forces in promoting the advance of forest into savanna.

Although these processes and their impacts may have eluded most researchers to date, they are commonplace to many of the region's rural inhabitants. And although researchers have been tempted to identify such techniques as recent innovations adopted in response to degradation, there is ample evidence to suggest that practices of this kind are ancient to many areas, albeit constantly refined in the light of changing circumstances and challenges, whether of an ecological or politico-economic kind.

Significantly, the vegetation forms that are the outcomes of these processes are also the key indicators from which ecologists and foresters have often deduced forest loss and degradation. Species-rich forest patches and isolated trees have been seen as the best representatives of a natural vegetation, now lost. But as we have shown, they can be interpreted very differently from a perspective which pays attention to local knowledge and history. Ironically, these supposed 'least disturbed' vegetation forms may actually be the most disturbed, testifying to people's capacity to enrich their landscape with desired vegetation. That there is evidence to support alternative interpretations for each of the countries we have studied here fundamentally questions the validity of such indicators of deforestation and, more generally, the methodology of interpreting landscape process from current vegetation form.

As we have now shown, this methodology has been a linchpin in the analysis of deforestation in each country, albeit used in varied combination with other methods and theories from natural and social sciences. For example, in Sierra Leone, deductions from vegetation form of this nature have been combined with deductions made from forestry exports and horse diseases. In Togo, deductions have been linked with presumptions concerning the impact of iron production. In Benin, observations of forest islands, palm forests and baobabs have been linked in support of particular interpretations of pre-colonial documents and of modern air photographs. In Ghana, Côte d'Ivoire, Benin and elsewhere, deductions from isolated trees and forest patches have been linked with assumptions concerning the full forest cover which once existed in climatically defined forest zones. In all the countries in question, deductions from vegetation form have been linked with views of demographic, social and economic history in which forests, once little disturbed, come under gradually increasing pressure. We are not arguing that this vision of forest change and its causes has no basis in reality. Clearly, there are times and places when such processes have operated. But we do argue that these processes have been exaggerated and overgeneralised, obscuring other processes and issues of major importance.

In this book, we have not been concerned with West Africa's dry forests, with the Sudanian and Sahelian woodlands lying to the north of our focal region. Nevertheless, scientific and policy concerns with the humid forest zone have always been linked, explicitly or implicitly, with those for these more northerly regions; for example, in ideas of all vegetation zones shifting southwards, or in the idea of the savannisation of the tropical forest being the 'outpost' of desertification processes. Indeed, the term 'desertification', as coined by Aubréville, was principally used to refer to the humid forest frontier.

Several of the themes which have underlain our reframing of humid forest history have been shown in recent work to be of equal relevance to the drylands. First, attempts to track environmental change through time series analysis, drawing on assorted historical sources, have provided an important challenge to conceptualising dryland degradation as a relentless one-way process. Vegetation in the Sahelian region has been found to fluctuate in accordance with rainfall variability, waxing and waning with the weather, rather than undergoing irrevocable decline (Mortimore 1989; Hellden 1991). Second, and even more than in forest ecology, ideas about dryland environmental change have been strongly influenced by non-equilibrium thinking, stressing the spatial and temporal variability of dryland vegetation at all scales, and the path dependency and hence historically contingent patterns of vegetation change (Holling 1973; Noy-Meir 1982; Ellis and Swift 1988; Westoby *et al.* 1989; Dublin *et al.* 1990; Behnke and Scoones

1993). This provides a radically different frame with which to investigate and describe grazing practices and their ecological impacts, and in doing so has 'rehabilitated' many pastoralist strategies which had been considered destructive and short-sighted and are now seen as logical and sustainable adaptations to dryland conditions (Sandford 1995; Scoones 1995). Third, the importance of management and settlement in transforming soil development and productivity has also been highlighted, suggesting how soils can be upgraded, whether by stimulating soil faunal activity or by importing organic fertilisers and materials (Mortimore *et al.* 1990).

Just as for the humid forest zone, evidence of landscape enrichment and the shaping of landscape by inhabitants should not be taken to suggest that Africa's drylands do not face ecological problems. Yet dryland evidence does contribute to a mounting critique of desertification as a concept with which to understand dryland ecological transformation, revealing it more as a politicised concept functional to particular institutions and power relations (Swift 1996).

The production of knowledge about forestry

Empirically, it would seem that the methods of assessing deforestation and the institutional structures of today's forestry and conservation circles seem to favour the production of exaggerated statistics of forest loss. It can be expected that our conclusions in this respect may also apply in regions outside the West African countries studied here which have a similar institutional and political history. Our results therefore suggest that data on tropical forest cover change during the twentieth century everywhere need to be examined more critically, and that this be treated as an urgent research priority.

Chapter 8 attempted to account for the tendency towards exaggerated views of deforestation in West Africa. The reasons, we have suggested, concern much more than scientific method in a narrow sense, and must be seen in historical context. Not only did the development of scientific ideas about West African forests have its own complex intellectual history and sociology, in which certain theories or debates were able to rise to the exclusion of others. But also, and crucially, these views dovetailed with the administrative and political concerns of the institutions with which they co-evolved in a process of mutual shaping. Ideas about forest–climate equilibria, or the functioning of relatively stable forest ecosystems, for instance, fed directly into a conceptual framework and set of scientific practices for conservation which was about external control, 'concerned precisely with establishing or recovering control, both over human impacts on nature (in "stopping habitat loss") and over nature itself (in habitat management)' (Adams 1997: 280). The institutions promoting conservation – government departments, and latterly also donor agencies and non-governmental organisations – have benefited from the instrumental effects of this knowledge in acquiring resource control from inhabitants. The scientific ideas which justified them are still felt almost tangibly in the location and laws of protected forest reserves, groves and trees.

At the same time, the linkages between scientific disciplines and policy have worked to simplify and homogenise forestry knowledge, and to deter conservation agencies further from addressing complex, unexpected social and ecological dynamics. When translated into policy documents, the detailed conclusions from field-based studies tend to be simplified into what Roe has termed development narratives: 'stories' which encode and stabilise the assumptions about a situation in such a way as to facilitate decision-making (Roe 1991). Narratives concerning forest change, the effects of farming or population–environment relations, for instance, suggest particular types of solution. They provide 'cultural scripts for action' which enable donors and government agencies to raise funds and act despite the uncertainties and complexities actually surrounding social and ecological change (see Hoben 1995; Leach and Mearns 1996). As others have argued, then, a tendency towards narrative making may be a longstanding and characteristic feature of the research–policy interface (see Hirschmann 1968).

It would appear from this study, however, that similar processes can work between scientific disciplines, and particularly between social and natural sciences. Analysts from each particular discipline might employ critical and detailed perspectives in their own research work, but often need to rely on simplifications from other disciplines to frame their broader research questions and trajectories. This helps account for the profusion of studies detailing the social dynamics of an assumed, unstudied deforestation process, for instance; or of the detailed botanical composition of forest–savanna boundaries assumed (in the absence of historical or climate history analysis) to be in retreat. Indeed, in this sense it could be argued that the strength of deforestation analysis lies in the interstices of disciplines. Thus, West African history and anthropology to date may have been shaped by erroneous vegetation history analysis which had its roots in colonial domination.

In this sense, this study exemplifies – in the particular context of West African forestry – the importance of taking a critical view of the production of knowledge in environment and development circles. It provides a regional case study in what has variously been characterised as the analysis of development discourse (e.g. Ferguson 1990; Grillo and Stirrat 1997; see Foucault 1980), or as post-structural or 'liberation' ecology (Escobar 1996; Peet and Watts 1996).

Implications for research into vegetation cover change

The arguments which we have forwarded concerning the long-term interactions between people and vegetation derive from a type of methodology rather different from normal ecological science. Central to our approach has been a privileging of historical data, in effect bringing history to the study of ecology, rather than ecology to the study of history. Rather than 'read' historical processes from landscape form, then, it has been the processes – and what can be known about them – which have largely driven the interpretation. Equally central to our approach has been to explore areas of dissonance between disciplinary perspectives, and to interrogate the mutually supportive dialogues between disciplines so often evident in deforestation analysis by reference to contradictory arguments forwarded in other literature, whether more recent, or long-present but ignored. To the extent that this can be done from secondary sources, we have emphasised the importance of paying serious attention to inhabitants' own knowledge and alternative perspectives or discourses on vegetation change.

Nevertheless, we have encountered many problems in using historical data and would not make a strong claim that those presented here add up to a 'hard', unequivocal, alternative story. In many cases, we have shown that it is possible for alternative, and strongly contrasting, interpretations to be made of the same 'facts'. In showing such ambiguity, we have opened up strong deductions to question but have not necessarily had the evidence necessary to support or reject them. There is value in showing that the data are questionable, and that major deductions to date have rested on very problematic bases. Our aim, then, has been to gain greater precision about vegetation change where, and insofar as, historical sources allow. Where precision is unachievable, our aim has been to raise and clarify the lines of debate when previously there has been none.

In many instances, steps towards further precision or the resolution of debate could only be taken through detailed field study. There are now a number of examples of the type of environmental history research that proves useful in such circumstances. Indeed, studies combining, to varying degrees, reconstructions of environmental time series through multiple data sources, with detailed analysis of social and political dynamics and attention to contrasting knowledges, seem to be a growing genre in the literature, with case studies (e.g. Tiffen *et al.* 1994; McCann 1995; Carswell *et al.* 1996) supported by general calls for such a methodological programme (e.g. Worster 1993; Peet and Watts 1996). In the West African forestry context, our country chapters have highlighted several key sites where such research programmes are very necessary. One is in the Togo Mountains, where work is much needed to put Guelly *et al.*'s intriguing findings concerning farm–fallow dynamics in historical and socio-political context. A second is in eastern Benin, where field research could usefully question Aubréville's and Chevalier's conclusions concerning the origin of isolated trees in farmers' fields.

Implications for conservation and development

As we have tried to show, within conservation circles at all dates deforestation has been portrayed as recent and the supposed period of rapid forest loss updated accordingly. This allows deforestation to be attributed to current generations and conveniently justifies a role for current state agencies and international organisations to intervene.

But as Table 9.1 underlines, the inhabitants of 15–20 million hectares of West Africa may be being misrepresented as having brought about deforestation during the twentieth century; in other words, within the lifetimes of themselves or their parents. Furthermore, if, as we have argued, the extent of forest vegetation in the forest–savanna transition zone has been increasing over the century, then a significant proportion of those blamed for deforestation may actually have experienced, and been partially responsible for, increased woody cover over their lifetimes. Gone unappreciated are the ways that many farmers have been enriching or stably managing their landscapes.

If our argument is correct, then the people who have been living in the area of the forest zone which has been only mythically deforested during this century are owed an apology. They have been blamed for damage which they have not caused and have paid heavily for this in policies aimed to control their so-called environmental 'vandalism', and to remove their control over resources in favour of national and international guardians (see Tanzidani 1993; Fairhead and Leach 1994a, 1995a; Leach and Mearns 1996). They are treated as strangers in their own, or hosts', ancestral lands, over which 'nature' and its national and international guardians have come to claim a right. Deforestation orthodoxies may, then, be adding unnecessarily to impoverishment, social upheaval and conflict in what is already a poor region. Inhabitants are denied not only their claims and control over valued resources, but also their own understandings of vegetation dynamics and the ecological and social histories with which these are entwined (see Amanor 1994a). Were one to add up the huge value of the timber and other resources alienated from inhabitants' control (and exploited and managed by others in the name of conservation and rational production), one would begin to appreciate an important element in processes of impoverishment in the region. Recall the events of 1852 in Sierra Leone when inhabitants of the Rosolo Creek region used their men's secret society to prohibit tree extraction by outsiders, leaving an estimated £16,000 worth of felled timber idle. Since then, such local institutions have become far less powerful in the face of the state and commercial interests which would exploit their resources.

As Chapter 8 argued, while the last few decades have seen many adjustments to forest policy and conservation approaches, often with the avowed aim of 'greater local participation', the conceptual framework of aid and policy, and their scientific substrate, have not fundamentally altered. Meanwhile the rise of global environmental concerns, and of global modelling as a practice and a profession, have created more, and more distanced, contexts in which deforestation figures are put to work, often with very little accountability to the people whose everyday lives may ultimately be influenced by them.

The analysis presented in this book is not an argument for less aid, or for a reduction of development effort. Equally, it does not deny the need to address major deforestation problems in particular places, whether linked to timber extraction, mining, commercial plantations or farming. Rather, it does suggest the urgent need for some changes of approach, some more practical, and others more conceptual and indirect. There are a number of practical, immediate implications for forest conservation. The first is the need to dismantle the dichotomy between the preservation of 'natural' or wild trees and the planting of new ones from nurseries, which pervades conservation and rural development thinking and practice. For cross-cutting this divide are farmers' many ways of working with and directing multiple ecological processes to encourage tree cover, whether involving active agency in the orientation of cultivation, wildling transplantation, soil or fire management, or the less intended effects of farming and settlement. There may well be practical techniques here around which farmers and foresters could collaborate when they share common goals in enriching a landscape with trees, although their application clearly requires close attention to the specificities of ecological and social relations. Our own work in Guinea attempted such an exercise (Fairhead and Leach 1996a).

A second and closely related implication concerns tenure over trees in fields, fallows and settlement sites. The outcomes of enrichment processes and historical legacies such as those considered above carry particular tenurial implications. These are often recognised in customary law, which has found ways to deal with the ambiguities presented. Yet they have been fundamentally neglected in statutory law, founded as this has generally been on colonially derived distinctions between planted and wild. Arguably, tree planting and community forest and land management programmes, which now give so much weight to the purported relationship between tenure and 'incentives', need to be founded on tree tenure concepts and categories which accord with local ecological and historical experience.

A third practical implication refers more to areas of currently forested land, which may be or become the subject of external conservation interest. Here, recognising the historical experiences of forest or forest-edge inhabitants would seem to be a prerequisite for any attempt to secure positive local involvement. Not only can past settlement or farming provide a basis for claims to land and resources that can, and will, crop up to override any newly imposed tenurial scheme, but also important are people's phenomenological experiences of forests as shaped by history: as an ex-social domain abandoned through the horrors of warfare, for instance. Indeed, the apparently 'spiritual' and 'sacred' attributes of forest imputed on the understandings of forest dwellers by conservationists keen to incorporate cultural sensitivity (see McNeely and Pitt 1985) may in many instances have less to do with attitudes to forest per se, or to categories of 'wilderness' in local thought, than to past events and inhabitation, and social memory of them. Clearly there are major implications here for reserve strategies and conservation-with-development, which would have to be dealt with on a case-by-case basis taking account of variations in history, administration and national legal frameworks. But in general, it would seem that a more positive approach to securing local support for forest protection would start by recognising tenurial claims and the social legacy of past events, and working out arrangements from this basis.

In many cases foresters on the ground have been forced to take such historically grounded claims into account owing to the political realities in which they work. The problem comes in articulating this recognition with the priorities set at national and above all international levels, which have a very different vision of forest status. For as we have seen, conceptions of uninhabited forest as 'nature', undisturbed by people, have provided tenurial grounds on the basis of vacancy for national and international guardians to intervene in habitat protection. Views of forest as an ecosystem at, or potentially at, equilibrium with climatic conditions in the absence of human disturbance have provided moral and scientific grounds for external management to override the 'disruptive' effect of local populations. Added to these are moral arguments based on the notions of forest areas as global commons or national patrimony, to be protected in an undisturbed state for a larger, future good.

The perspectives of new, more historically grounded ecological thinking, like the arguments forwarded in this book, remove such baselines against which to assess forest loss or deterioration. In this, they raise some profound implications for the conceptual framework of forest policy and conservation. As Adams emphasises, 'Gone are the days when ecologists (and conservationists) could conceive of "nature" in equilibrium, and hence portray human-induced change in those ecosystems as somehow "unnatural". . . . Human actions are part of the web of influences on ecological change, not external equilibrium-disturbing impacts' (1997). And as McNeely summarises:

> because chance factors, human influence and small climatic variation can cause very substantial changes in vegetation, the biodiversity for any given landscape will vary substantially over any significant time period – and no one variant is necessarily more 'natural' than the others.
>
> (McNeely 1994: 3)

In this sense, new ecological thinking undermines the basic principles on which conservation action has been built – the control of human impacts. It could be argued that it lets in, instead, a laissez-faire approach in which, in the absence of any secure basis for judging change, 'anything goes'. Recognising this dilemma, Worster (1994: 3) concludes that not all changes are equal, and some are more acceptable than others (see Botkin 1990: 157). But this is to beg the question of who judges acceptability. Arguably, conventional approaches to conservation – while grounded in different ecological theories – have been similarly based on certain agents (e.g. government forestry departments) allowing certain types of 'acceptable' change (e.g. succession to climax) while restricting others (such as forest conversion to fallow). We would suggest, instead, that recent ecological perspectives put ecology and conservation even more firmly than before into the domain of politics. As McNeely (1994) argues, with no 'baseline nature', conservation becomes very clearly a question of social or political choice about what vegetation or biodiversity forms are desirable at any given time in social history. In conceptual terms, at least, space is opened up for debate and negotiation among different users as to what constitutes value in vegetation, and which values should take priority; and putting such concepts into action comes to demand participatory, or 'democratic', forms of environmental politics.

A fundamental requirement for pursuing the above implications in any particular place is to have clarity about the history of forest cover change. This is a necessary first step to revealing the ecological 'rules of the game', and the types of contingency and path dependency which have been operating and are likely to be important. It is, furthermore, necessary to reveal local landscape enrichment practices which may have been obscured by ahistorical analyses. West African people may have been working with ecological dynamics far more effectively than has been imagined. Rather than seek to condemn and amend their practices, conservation grounded in a reworked, historicised ecology may have much to learn from them. Historical analysis is necessary to reveal the existence of – if perhaps not to clarify the precise nature of – tenurial claims based on past events and situations.

Any fundamental shift in conservation, then, will require changes in the types of research that inform policy. Bringing historians – hitherto largely neglected by forest departments, yet often possessing a wealth of area-specific knowledge and methodological skill – into the policy research process would be a step in the right direction. So too would a greater use of research methodologies geared to picking up on social and ecological dynamics. An accumulation of research from historical and local perspectives, suggesting different local solutions, is clearly necessary to push the agendas for particular countries. This suggests a need for support to research both in West African universities, and undertaken in a policy context, which uses methods open enough to allow for a reconceptualisation of trajectories of change in particular places.

Yet analysis here suggests that the problem may be more deeply rooted and intractable, and that such calls for change are naive. This book has examined six countries, each with different colonial and administrative histories, different ecologies and different patterns of social and economic change. Yet in each, dominant analyses of deforestation proved to be strongly exaggerated, and to be grounded in strikingly similar misconceptions. In seeking out the scientific and institutional reasons for this, we have suggested that exaggerated statistics were merely the surface of something much deeper. The earlier paradigm of ecological stability has generated an edifice of analysis not only in forestry but also in other disciplines; an edifice intimately linked with the development of forestry and conservation institutions, and with the control of valuable forest land and resources. For it is undeniably the case that whether through extraction or conservation, forests and their resources generate vast revenues. In this context, changes in policy and donor funding approach will be difficult and inevitably meet with resistance. The way forward is therefore problematic.

To provide a context into which emergent local analyses might fit and find their place, there is clearly a need to break the endless cycle of flawed regional statistics for deforestation which circulate internationally. In part, this is necessary to alter the images of forests portrayed by conservation organisations in the North; images grounded in the northern cultural precepts and historical experience as well as the ecological science with which these are linked (see Peet and Watts 1996). Changes in such images are necessary to legitimate a shift in practical approaches on the ground; and to legitimate – rather than extinguish or minimise – the findings of local studies.

Such far-reaching shifts call not only for changes in the types of research which might inform policy but, more fundamentally, for reconceptualisation of the research–policy process and its power relations. Such calls, voiced in various ways, are evident in some recent work on environment and development. Peet and Watts (1996) emphasise the need to reveal alternative discourses and put them into circulation; and to counterpose, in an explicit way, the 'environmental imaginaries', or historically situated, naturally informed knowledges of different groups. Escobar (1995, 1996), in a similar vein, particularly emphasises the role of new social movements. Others have phrased related concerns in terms of a call for participatory research, assisting a reconceptualisation of problematics in multiple local situations, so as to bring about, cumulatively and over time, reversals in 'normal professionalism' (Pimbert and Pretty 1995; Chambers 1996). Work on science and public policy argues for an approach to policy research – especially in complex, uncertain environmental situations – which renders explicit plural 'rationalities' and their underlying social and political commitments, and brings them into a common arena for debate, negotiation and consensus building (Wynne 1992; Thompson 1993; see Leach and Mearns 1996).

In their different ways, these suggestions all assume that changes in global processes will result from a percolation upwards of local findings; from what Funtowicz and Ravetz (1992) term a 'democratisation of expertise'. While endorsing their spirit, we nevertheless end with an additional and more direct call – a plea, if you like – to those who are already loci of power in the production of knowledge. International analysts, organisations and modellers must recognise their responsibility in generating and circulating forestry knowledge. On the one hand, they must seek greater rigour in their data and interpretation, linking up with those – local academics, perhaps – who can assist in this. On the other hand, rather than continuing to work uncritically with existing data in the absence of anything better, they must be more critically aware of their origins, biases and possible flaws. We have shown that the forestry statistics in international circulation are the epiphenomena of power relations with long historical roots. Their reiteration is far from neutral, but serves to reinforce those power relations in ways, and with effects, from which their proponents might prefer to be dissociated.

Note

1 This analytical perspective was not entirely ignored by early botanists, as the works of Chevalier (1912b) and Aubréville (1938) testify.

BIBLIOGRAPHY

Abimbola, W., 1964, 'The ruins of Oyo Division', *African Notes* 2(1): 16–19. Institute of African Studies, Ibadan, Nigeria.

Abraham, A., 1978, *Mende Government and Politics under Colonial Rule: A Historical Study of Political Change in Sierra Leone 1890–1937*. Freetown: Sierra Leone University Press.

Adam, J.G., 1948, 'Les reliques boisées et les essences des savanes dans la zone préforestière en Guinée française', *Bulletin de la Société Botanique Française* 98: 22–26.

—— 1949, 'La végétation de la région de la source du Niger' (Conférence Africaine des Sols 1948, Goma), *Bulletin Agric. Congo Belge* 40(2): 1363–1374.

—— 1968, 'Flore et végétation de la lisière de la forêt dense en Guinée', *Bulletin d'IFAN* series A, 30(3): 920–952.

Adam, J.G., 1969, 'Etude comparée de quelques forêts ouest-africanes (Sierra Leone et Libéria)', Bulletin de l'IFAN, t. xxxi, Série A.

Adams, J.S. and T.O. McShane, 1992, *The Myth of Wild Africa: Conservation without Illusion.* New York and London: W.W. Norton.

Adams, W.M., 1997, 'Rationalisation and conservation: ecology and the management of nature in the UK', *Transactions of the Institute of British Geographers* 22: 277–291.

Adejuwon, J.O. and F.A. Adesina, 1992, 'The nature and dynamics of the forest–savanna boundary in south-western Nigeria', in P.A. Furley *et al.* (eds) *Nature and Dynamics of Forest–Savanna Boundaries*, 331–352. London: Chapman and Hall.

Adjanohoun, E., 1964, 'Végétation des savanes et des rochers découverts en Côte d'Ivoire centrale', *Mémoire ORSTOM* 7, Paris.

—— 1966, 'Conservation de la végétation et de ses espèces au Dahomey'. *Etudes Dahoméennes* 8: 29–38.

Adjanohoun, E. and A.L. Assi, 1968, 'Essais de création de savanes incluses en Côte d'Ivoire forestière', Université d'Abidjan, *Annales de la Faculté des Sciences* 4: 237–256.

Adu-Boahen, A., 1977, 'Ghana before the coming of Europeans', *Ghana Social Science Journal* 4(2): 93–106.

Ahin, K., 1973, 'Asante security posts in the northwest', seminar on Asante and Brong, Institute of African Studies, University of Ghana, March 1973.

BIBLIOGRAPHY

Ahn, P.M., 1959, 'The principal areas of remaining original forest in western Ghana and their agricultural potential', Ghana Division of Agriculture, Kumasi.

Akpan, M.B., 1982–3, 'Native administration and Gola-Bandi resistance in north-western Liberia 1905–1919', *Journal of the Historical Society of Nigeria* 11(3–4): 142–164.

Alba, A., 1956, 'Le développement de la foresterie en Afrique-Occidentale française', *Journal of the West African Science Association* 2: 158–171.

Albeca, A., 1894, L'Avenir du Dahomey', *Annales de Géographie* 4(1894–5): 166–221.

Aldinger, C., 1867, 'Das Kroboland', *Der evangelische Heidboten* 6.

Alldridge, T.J., 1901, *The Sherbro and its Hinterland.* London: Macmillan.

—— 1910, *A Transformed Colony: Sierra Leone as It Was, and as It Is, its Progress, Peoples, Native Customs and Undeveloped Wealth.* London: Seeley.

Allen, J.C. and D.F. Barnes, 1985, 'The causes of deforestation in developing countries', *Annals of the Association of American Geographers* 75(2): 163–184.

Allison, P.A., 1962, 'Historical inferences to be drawn from the effect of human settlement on the vegetation of Africa', *Journal of African History* 3(2): 241–249.

Almada, Andre Alvares de, 1984 [1594], *Brief Treatise on the Rivers of Guinea. Being an English Translation of a Variorum Text, organised by the late A.T. da Mota.* Translation by P.E.H. Hair and notes by P.E.H. Hair and J. Boulegue. Unpublished ms; Liverpool.

Alvares, Manuel, c. 1615, *Ethiopia Minor and a Geographical Account of the Province of Sierra Leone.* Transcription from an unpublished manuscript by the late A.T. da Mota and Luis de Matos. Translation and introduction by P.E.H. Hair. Unpublished ms; Liverpool.

Amanor, K.S., 1993, Wenchi Farmer Training Project: Social/Environment Baseline Study. Accra: unpublished report to Overseas Development Administration.

Amanor, K.S., 1994a, *The New Frontier: Farmer Responses to Land Degradation.* London: UNRISD and Zed Books.

—— 1994b, 'Farmer experimentation and changing fallow ecology in the Krobo district of Ghana', in W. de Boef, K. Amanor and K. Wellard (eds) *Cultivating Knowledge: Genetic Diversity, Farmer Experimentation and Crop Research*, 35–43. London: Intermediate Technology Publications.

Andah, B.W., 1988, 'The peoples of Upper Guinea (between the Ivory Coast and the Casamance)', in M. El Fasi (ed.) *General History of Africa III: Africa from the Seventh to the Eleventh Century*, 530–558. UNESCO California: Heinemann.

—— 1992, 'Identifying early farming traditions of West Africa', in T. Shaw, P. Sinclair, B. Andah and A. Okpoko (eds), *The Archaeology of Africa: Food, Metals and Towns.* London: Routledge.

Anderson, A.B. and D.A. Posey, 1989, 'Management of a tropical scrub savanna by the Gorotire Kayapo of Brazil', *Advances in Economic Botany* 7: 159–173.

Anderson, B., 1870, *Narrative of a Journey to Musardu, Capital of the Western Mandingoes.* New York: S.W. Green.

—— 1874 (1912), *Narrative of the Expedition Despatched to Musahdu by the Liberian Government under Benjamin K. Anderson, Senior in 1874.* Ed. Frederik Starr. Monrovia College of West Africa Press, Liberia.

BIBLIOGRAPHY

Anderson, D. and R. Grove, 1987, *Conservation in Africa: People, Policies and Practice.* Cambridge: Cambridge University Press.

Anderson, D.M., 1984, 'Depression, dust bowl, demography, and drought: the colonial state and soil conservation in East Africa during the 1930s', *African Affairs* 83(332): 321–343.

Anquandah, J., 1982, *Rediscovering Ghana's Past*, Harlow: Longman.

Anquandah, J., 1992, 'The Kintampo complex: a case study of early sedentism and food production in sub-Sahelian West Africa', in T. Shaw, P. Sinclair, B. Andah and A. Okpoko (eds) *The Archaeology of Africa: Food, Metals and Towns.* London: Routledge.

Anyadike, R.N.C., 1982, 'Natural vegetation in relation to the moisture situation in West Africa', *Bulletin d'IFAN* series A, 44(3–4): 221–233.

AOF (Afrique-Occidentale française), 1934, *Annuaire statistique de l'AOF*. Paris.

Apter, D.E., 1963, *Ghana in Transition.* New York: Atheneum.

Arnaud, J.-C. and G. Sournia, 1979, 'Les forêts de Côte d'Ivoire: une richesse naturelle en voie de disparition', *Cahiers d'Outre-Mer* 127: 281–301.

—— 1980, 'Les forêts de Côte d'Ivoire', *Annales de l'Université d'Abidjan* series G, 9: 6–93.

Asare, E.O., 1962, 'A note on the vegetation of the transition zone of the Tain Basin in Ghana', *Ghana Journal of Science* 2: 60–373.

Atherton, J.H., 1972, 'Excavations at Kamabai and Yagala rock shelters, Sierra Leone', *West African Journal of Archaeology* 2: 39–74.

Aubréville, A., 1932, 'La forêt de la Côte d'Ivoire: essai de géobotanique forestière', *Bulletin commercial et historique du science, AOF* 15(2–3): 205–249.

—— 1937, 'Les forêts du Dahomey et du Togo', *Bull. du Com. d'Etud. Hist et Scient. de l'A.O.F.* 20(1–2): 1–221.

—— 1938, 'La forêt coloniale: les forêts de l'Afrique-Occidentale française', *Annales d'Académie des Sciences Coloniales* IX. Paris: Société d'Editions Géographiques, Maritimes et Coloniales.

—— 1939, 'Forêts reliques en Afrique-Occidentale française', *Revue de Botanique Appliquée et Agronomie Tropicale* 215: 479–484.

—— 1947a, 'Les brousses secondaires en Afrique-Equatoriale', *Bois et Forêts des Tropiques* 1 (1947): 29–49.

—— 1947b, 'Erosion et bovalisation en Afrique noire française', *L'Agronomie Tropicale* 2(1947): 24–35.

—— 1949 *Climats, forêts et désertification de l'Afrique tropicale.* Paris: Société d'Edition de Géographie Maritime et Coloniale.

—— 1956, 'Tropical Africa', in S. Haden-Guest, J.K. Wright and E.M. Teclaff (eds) *A World Geography of Forest Resources*, 353–384. New York: Ronald Press.

—— 1957, 'A la recherche de la forêt en Côte d'Ivoire', *Bois et Forêts des Tropiques* 56 and 57, Nogent sur Marne (France).

—— 1959, *La flore forestière de la Côte d'Ivoire.* Publ. CTFT no. 15 (3 vols), Nogent-sur-Marne (France)

—— 1962, 'Savanisation tropicale et glaciation quaternaire', *Adansonia* II(1): 233–237.

—— 1966, 'Les lisières forêt-savane dans les régions tropicales', *Adansonia* VI(2): 175–187.

BIBLIOGRAPHY

Avenard, J.-M., 1968, *Réflexions sur l'état de la recherche concernant les problèmes posés par les contacts forêts–savanes*. Paris: ORSTOM.

Avenard, J.-M., J. Bonvallot, M. Latham, M. Renard-Dugerdil and J. Richard, 1974, *Aspects du contact forêt–savane dans le centre et l'ouest de la Côte d'Ivoire: étude descriptive*. ORSTOM: Abidjan.

Baker, H.G., 1962, 'Comments on the thesis that there was a major centre of plant domestication near the headwaters of the river Niger', *Journal of African History* III(2): 229–233.

Bakshi, T.S., 1963, 'Vegetation', in A.R. Stobbs (ed.) *The Soils and Geography of the Boliland region of Sierra Leone*. Freetown: Government of Sierra Leone.

Balée, W. 1989, 'The culture of Amazonian forests', *Advances in Economic Botany* 7: 1–21.

Barber, R.J., 1985, 'Land snails and past environment at the Igbo-Iwoto Esie site, southwestern Nigeria', *West African Journal of Archaeology* 985: 89–102.

Barbot, J., 1678–9, 'Journal d'un voyage de traité en Guinée, à Cayenne et aux Antilles fait par Jean Barbot en 1678–1679', G. Debien, M. Delafosse and G. Thilmans (eds) *Bulletin d'IFAN* 40, series B2 (1978): 235–395.

Barnes, R.F.W., 1990, 'Deforestation trends in tropical Africa', *African Journal of Ecology* 28: 161–173.

Barraclough, S.L. and K. Ghimire, 1990, 'The social dynamics of deforestation in developing countries: principal issues and research priorities', *UNRISD Discussion Paper* 16, Geneva.

—— 1995, *Forests and Livelihoods: The Social Dynamics of Deforestation in Developing Countries*. UNRISD and Macmillan.

Barros, P. de, 1986, 'Bassar: a quantified, chronologically controlled, regional approach to a traditional iron production centre in West Africa', *Africa* 56(2): 148–173.

—— 1988, 'Societal repercussions of the rise of large-scale traditional iron production: a West African example', *The African Archaeological Review* 6: 91–113.

Begue, L. 1937, 'Contribution à l'étude de la végétation forestière de la haute Côte d'Ivoire', *Cte. d'Et. Hist. et Scient. de l'AOF* series B, 4.

Behnke, R. and I. Scoones, 1993, 'Rethinking range ecology: implications for rangeland management in East Africa', in R.H. Behnke, I. Scoones and C. Kerven (eds) *Range Ecology at Disequilibrium*, 1–30. London: ODI.

Beinart, W., 1984, 'Soil erosion, conservationism and ideas about development: a southern African exploration, 1900–1960', *Journal of Southern African Studies* 11(1): 52–83.

—— 1996, 'Soil erosion, animals and pasture over the longer term: environmental destruction in Southern Africa', in M. Leach and R. Mearns (eds) *The Lie of the Land: Challenging Received Wisdom on the African Environment*, 54–72. Oxford: James Currey and Portsmouth: Heinemann.

Bellier, A., D. Gillon, Y. Gillon, J.-L. Guillaumet and A. Perraud, 1969, 'Recherches sur l'origine d'une savane incluse dans le bloc forestier du Bas Cavally (Côte d'Ivoire) par l'étude des sols et de la biocenose', *Cahiers ORSTOM*, series Biology 4: 65–95.

Bertrand, A., 1983, 'La déforestation en zone de forêt en Côte d'Ivoire', *Bois et Forêts des Tropiques*, 202: 3–17.

Bilsborrow, R. and M. Geores, 1994, 'Population, land-use and the environment in developing countries: what can we learn from cross-national data?', in K. Brown and D.W. Pearce (eds) *The Causes of Tropical Deforestation: The Economic and Statistical Analysis of Factors Giving Rise to the Loss of Tropical Forests*. London: UCL Press.

Blanc-Pamard, C., 1978, 'Contraintes écologiques et réalités socio-économiques', *Cahiers ORSTOM*, series Sciences Humaines, 15(1): 51–79.

—— 1979, 'Un jeu écologique différentiel: les communautés rurales du contact forêt-savane au fond du "V Baoulé"', *Travaux et Documents de l'ORSTOM* 107. Paris: ORSTOM.

Blanc-Pamard, C. and P. Peltre, 1987, 'Remarques à propos de "Ecologies et histoire: les origines de la savane du Bénin"', *Cahiers d'Etudes Africaines* 107–108 (XXVII-3–4): 419–423.

Blyden, E.W., 1869, 'The Boporo country – Monrovia to Boporo', in W.A. Givens (ed. 1966) *Selected Works of Dr. Edward Wilmot Blyden*, 277–295; 303–305. Robertsport, Liberia: Tubman Center of African Culture.

—— 1872, 'Report on the expedition to Falaba', *Journal of the Royal Geographical Society* 17(2).

—— 1873, 'Report of the expedition to Timbo, January–March 1873', in H.R. Lynch (ed. 1978) *Selected Letters of Edward Wilmot Blyden*. New York: Kto Press.

Boateng, E.A., 1962, 'Land-use and population in the forest zone of Ghana', *Bulletin of the Ghana Geographical Association* 7: 14–20.

Boserup, E., 1965, *The Conditions of Agricultural Growth: The Economics of Agrarian Change under Population Pressure.* Republished 1993, London: Earthscan Publications.

Bosman, W., 1705 (1967), *A New and Accurate Description of the Coast of Guinea.* London: Frank Cass (reprint of 1705 1st edn with introduction by J.R. Willis).

Botkin, D.B., 1990, *Discordant Harmonies: A New Ecology for the Twenty-first Century*. New York: Oxford University Press.

Bourret, F.M., 1960, *Ghana: The Road to Independence.* Oxford: Oxford University Press.

Bousquet, B., 1978, 'Un parc de forêt dense en Afrique: le Parc National de Taï (Côte d'Ivoire)', *Bois et Forêts des Tropiques* (May–June): 27–46, 180; (July–August): 23–37.

Bowdich, T.E., 1819, *A Mission from Cape Coast Castle to Ashantee.* London: John Murray.

Brand, R.B., 1976, 'Les hommes et les plantes: l'usage des plantes chez le wemenou du sud-Dahomey', *Geneve Afrique* 15(1): 15–44.

Breemer, J.P.M. van den, 1992, 'Ideas and usage: environment in Aouan society, Ivory Coast', in E. Croll and D. Parkin (eds) *Bush Base: Forest Farm: Culture, Environment and Development*, 97–109. London: Routledge.

Brenan, J.P.M., 1978, 'Some aspects of the phytogeography of Tropical Africa', *Ann. Missouri Bot. Gard.* 65: 437–478.

Breschin, A., 1902, 'La forêt tropicale en Afrique, principalement dans les colonies françaises', *La Géographie* 5: 431–450, 6: 27–39.

Brooks, G.E., 1985, 'Western Africa to c. 1860 AD. A provisional historical schema based on climatic periods', Indiana University African Studies Programme Working Paper Series no. 1.

—— 1986, 'A provisional historical schema for Western Africa based on seven climatic periods', *Cahiers d'Etudes Africaines* 101–102(XXVI-1-2): 43–62.

—— 1993, *Landlords and Strangers: Ecology, Society and Trade in Western Africa, 1000–1630*. Boulder and Oxford: Westview Press.

Brou-Tanoh, A., 1967, 'La tradition orale chez les Agni-Ahali du Moronou', *Bulletin d'Information et de Liaison des Institutes d'Ethnosociologie et de Géographie Tropicale* (Abidjan: University of Abidjan) 2: 45–48.

Brouwers, J.H.A.M., 1993, *Rural People's Response to Soil Fertility Decline: The Adja Case (Benin)*. Wageningen: Agricultural University Wageningen.

Brown, K., 1993a, 'Economics and the conservation of global biological diversity', Global Environmental Facility Working Paper 2, Washington, DC.

—— 1993b, 'Global environmental change and mechanisms for North–South resource transfer', *Journal of International Development* 5(6): 571–589.

Brown, K. and D.W. Pearce (eds), 1994, *The Causes of Tropical Deforestation: The Economic and Statistical Analysis of Factors Giving Rise to the Loss of Tropical Forests*. London: UCL Press.

Brown, S., 1993, 'Tropical forests and the global carbon cycle: the need for sustainable land-use patterns', *Agriculture, Ecosystems and Environment* 46: 31–44.

Brown, S., J. Sathaye, M. Cannell and P.E. Kauppi, 1996, 'Management of forests for mitigation of greenhouse gas emissions', in R.T. Watson, M.C. Zinyowera, R.H. Moss and D.J. Dokken *Climate Change 1995 – Impacts, Adaptations and Mitigation of Climate Change: Scientific Technical Analyses*, 773–798. Cambridge: Cambridge University Press. Published for the Intergovernmental Panel on Climate Change.

Brownell, T.C., 1869, 'Missionary exploration by a native', *African Repository* 45: 308–313.

Bruce, J.W., 1990, 'Community forestry: rapid appraisal of land and tree tenure', Community Forestry Note 5. Rome: FAO.

Buechner, H.K. and H.C. Dawkins, 1961, 'Vegetation change induced by elephants and fire in Murchison Falls National Park, Uganda, *Ecology* 42: 752–766.

Burgess, J.C., 1992, 'Economic analysis of the causes of tropical deforestation', Discussion Paper 92–03, London Environmental Economics Centre.

Bürgi, E., 1888, 'Reisen an der Togoküste und im Ewegeiet', *Petermanns Mitteilungen* 34(8): 233–237.

Burton, R.F., 1864 (1966), *A Mission to Gelele, King of Dahome*. London: Routledge and Kegan Paul.

Buxton, P.A., 1955, 'The natural history of tsetse flies', *London School of Hygiene and Tropical Medicine Memoir* 10: 261–314.

CAB International (Development Services), n.d., *Deforestation in Africa: A Literature Review*.

Carswell, G. *et al.*, 1996, 'More people, more fallow: environmental change in Uganda', unpublished report.

Chambers, R., 1996, *Whose Reality Counts? Putting the First Last*. London: IT Publications.

Charter, C., 1946, *Annual Report of the Soil Scientist*. Ghana: West African Cacao Research Institute.

Charter, J.R. and R.W.J. Keay, 1960, 'Assessment of the Olokomeji fire-control experiment (investigation 254) 28 years after institution', *Nigerian Forestry Information Bulletin* 3: 1–32.

Chauveau, J.P., J.P. Dozon and J. Richard, 1981, 'Histoires de riz, histoires d'Igname: le cas de la moyenne Côte d'Ivoire', *Africa* 51(2): 621–657.

Chevalier, A., 1900, 'Les zones et les provinces botaniques de l'Afrique-Occidentale française', *Comptes Rendus des Séances de l'Académie des Sciences* 130(18): 1202–1208.

—— 1909a, 'L'extension et la regression de la forêt vierge de l'Afrique tropicale', *Comptes Rendus des Séances de l'Académie des Sciences*, séance 30 August: 458–461.

—— 1909b, 'Les massifs montagneux du nord-ouest de la Côte d'Ivoire', *La Géographie* 20: 207–224.

—— 1909c, 'Les végétaux utiles de l'Afrique tropicale française: première étude sur les bois de la Côte d'Ivoire', *Etudes Scientifiques et Agronomiques*, Fascicule V.

—— 1910, 'Le pays des Hollis et les regions voisines', *La Géographie* 21.

—— 1911, 'Essai d'une carte botanique forestière et pastorale de l'A.O.F.', *Comptes Rendus de l'Académie des Sciences* CLII(6), June.

—— 1912a, 'Carte botanique, forestière et pastorale de l'Afrique-Occidentale française', *La Géographie* 26(4).

—— 1912b, *Rapport sur une mission scientifique dans l'ouest africain (1908–1910)*. Paris: Missions Scientifiques.

—— 1920, *Exploration botanique de l'A.O.F.* Paris.

—— 'Sur la dégradation des sols tropicaux causée par les feux de brousse et sur les formations végétales régressives qui en sont la conséquence', *Comptes Rendus de l'Académie des Sciences* CLXXXVIII: 84–86.

—— 1933a, 'Les bois sacrés des Noirs de l'Afrique tropicale comme sanctuaires de la nature', *C. R. Société de Biogéographie* 1933: 37.

—— 1933b, 'Le territoire géobotanique de l'Afrique tropicale nord-occidentale et ses subdivisions', *Bulletin de la Société Botanique de France*, séance 13 January.

—— 1948a, 'Biogéographie et écologie de la forêt dense ombrophille de la Côte d'Ivoire', *Rev. Int. de Bot. appl. et d'Agr. trop.* 28: 101–115.

—— 1948b, 'L'Origine de la forêt de la Côte d'Ivoire', *C. R. Société de Biogéographie*.

Chevalier, A. and D. Normand, 1946, *Forêts vierges et bois coloniaux*. Paris: P.U.F. Coll. Que sais je? no. 143.

Chevalier, A. and M. Perrot, 1911, *Les kolatiers et les noix de kola*. Paris: Challamel.

Chipp, T., 1922a, 'Danger from forest destruction on the Ejura and Mampong scarps', *Journal of the Gold Coast Agricultural and Commercial Society* March–June, 3: 131–137.

—— 1922b, *The Forest Officer's Handbook of the Gold Coast, Ashanti and Northern Territories*. London: Crown Agents.

—— 1927, 'The Gold Coast forest: a study in synecology', *Oxford Forestry Memoires* 7.

Church, R.J.H., 1968 (6th edn), *West Africa: A Study of the Environment and of Man's Use of It*. London: Longmans.

Church Missionary Intelligence, 1870, 'Journal notes by the Rev. A. Menzies of an expedition to the Mende country, with a missionary of the American Society, 1869', vol. 6, new series: 84–96.

Clayton, W.D., 1958, 'Secondary vegetation and the transition to savanna near Ibadan, Nigeria', *Journal of Ecology* 46: 217–238.

—— 1961, 'Derived savanna in Kabba Province, Nigeria', *Journal of Ecology* 46: 595–604.

Cleaver, K., 1992, 'Deforestation in the western and central African forest: the agricultural and demographic causes, and some solutions', in K. Cleaver *et al.* (eds) *Conservation of West and Central African Rainforests*, 65–78. World Bank Environment Paper no. 1. Washington, DC: World Bank.

Clements, F.E., 1916, 'Plant succession: an analysis of the development of vegetation', Carnegie Institute Washington Publication, 242: 1–512.

Coelho, F. de L., 1684, *Description of the Coast of Guinea*. Introduction and English translation of the Portuguese text by P.E.H. Hair. October 1985.

Cole, N.H.A., 1968, *The Vegetation of Sierra Leone*. Sierra Leone: Njala University Press.

—— 1980, 'The Gola forest in Sierra Leone: a remnant of tropical primary forest in need of conservation', *Environmental Conservation* 7(1): 33–40.

Cooper, J.P., 1931, 'The evergreen forests of Liberia', *Yale University Bulletin* 31.

Cornevin, R., 1959 (1st edn), 1969 (3rd edn), *Histoire du Togo*. Paris: Editions Berger-Levrault.

Croft, J.A., 1874, 'Exploration of the River Volta, West Africa', *Proceedings of the Royal Geographical Society* 18(I–V): 183–194.

Crone, G.R. (ed.), 1937, *The Voyages of Cadamosto*. London: Hakluyt Society.

CTFT (Centre Technique Forestier Tropical), 1966, 'Ressources forestières et marché du bois en Côte d'Ivoire', Abidjan: CTFT/SODEFOR.

Curtin, P., 1969, *The Atlantic Slave Trade: A Census*. Madison: University of Wisconsin Press.

Dale, I.R., 'Forest spread and climatic change in Uganda during the Christian era', *Empire Forestry Journal* 33: 23–29.

Dalziel, J.M., 1937, *The Useful Plants of West Tropical Africa*. London: Crown Agents.

Dapper, O., 1668, *Naukeurige beschrijvinge der Afrikaensche gewesten* . . . Amsterdam.

—— 1670 (1967), Translation of section related to Ghana, *Ghana Notes and Queries* 1967: 16.

Daveau, S., 1963, 'The Loma mountains', *The Bulletin: The Journal of the Sierra Leone Geographical Association* 9: 2–11.

Davidson, J., 1971, 'Trade and politics in the Sherbro hinterland, 1849–1890', PhD dissertation, University of Wisconsin.

Davies, G. and P. Richards, 1991, *Rain Forest in Mende Life: Resources and Subsistence Strategies in Communities around Gola North Forest Reserve*. Report to ESCOR of the Overseas Development Administration, London.

Davies, O., 1964, 'Archaeological exploration in the Volta Basin', *Bulletin of the Ghana Geographical Association* 9(2): 28–33.

—— 1967, *West Africa before the Europeans: Archaeology and Prehistory*. London: Methuen.

D'Azevedo, W.L., 1962a, 'Some historical problems in the delineation of a Central West Atlantic Region', *Annals, New York Academy of Sciences* 96: 513–538.

—— 1962b, 'Uses of the past in Gola discourse', *Journal of African History* 11: 11–34.

Debrunner, H., 1970, 'Notes sur les peuples témoins du Togo, à propos de sites montagneux abandonnés', *Bulletin d'Enseignement Supérieur du Bénin*, 11 (Oct–Nov 1969), 12 (Feb–March 1970) and 13 (May–June 1970).

De Faro, A., 1663–4, 'Andre de Faro's missionary journey to Sierra Leone in 1663–4. A shortened version, in English translation'. Ed. P.E.H. Hair. Institute of Africa Studies, University of Sierra Leone Occasional Paper no. 5. 1982.

Delafosse, M., 1908, *Les frontières de la Côte d'Ivoire*. Paris: Masson.

Deluz, A., 1970, *Organisation sociale et tradition orale: les Gouro de Côte d'Ivoire*. Paris: Mouton.

Department of Soil and Land Use Survey, Ghana (1956) *Report on the Department of Soil and Land-Use Survey, 1956*, Accra: Government Printer.

Dewees, P., 1989, 'The woodfuel crisis reconsidered: observations on the dynamics of abundance and scarcity', *World Development* 17(8): 1159–1172.

Dickson, K.B., 1971, *A Historical Geography of Ghana*. Cambridge: Cambridge University Press.

Diop, L.-M., 1978, 'Le sous-peuplement de l'Afrique noire', *Bulletin de l'IFAN* 40B(4): 718.

Dixon, R.K., S. Brown, R.A. Houghton, A.M. Solomon, M.C. Trexler and J. Wisniewski, 1994, 'Carbon pools and flux of global forest ecosystems', *Science* 263, 14 January: 185–190.

Djegui, N., 1995, 'Les stocks organiques dans les sols cultivés sous palmeraie et cultures vivrières dans le sud du Bénin', in *Fertilité du milieu et stratégies paysannes sous les tropiques humides*, 189–193. Report of Seminar, 13–17 November 1995, Montpellier. Montpellier: CIRAD.

Donelha, A., 1625, *An Account of Sierra Leone and the Rivers of Guinea of Cape Verde*. Edition of Portuguese text, introduction and notes and appndices by A.T. da Mota, notes and English translation by P.E.H. Hair. Lisbon: Centro de Estudos de Cartografia Antiga, 19. 1977.

Donno, E., 1976, *An Elizabethan in 1582. The Diary of Richard Madox, Fellow of All Souls*. London: The Hakluyt Society.

Dorm-Adzobu, C., 1974, 'The impact of migrant Ewe cocoa farmers in Buem, the Volta region of Ghana', *Bulletin of the Ghana Geographical Association* 16: 45–53.

—— 1985, 'Forestry and forest industries in Liberia. An example in ecological destabilization', *Internat. Inst. für Umwelt und Gesellschaft* pre-85, 14, Berlin.

Dorward, D.C. and A.I. Payne, 1975, 'Deforestation, the decline of the horse and the spread of the tsetse fly and trypanosomiasis (Nagana) in nineteenth century Sierra Leone', *Journal of African History* 16(2): 241–256.

Dublin, H., A. Sinclair and J. McGlade, 1990, 'Elephants and fire as causes of multiple stable states in the Serengeti-Mara woodlands', *Journal of Animal Ecology* 59: 1147–1164.

Duncan, J., 1847, *Travels in Western Africa, in 1845 & 1846*. London.

Dunn, R.M., 1989, *Forest Inventory Project: End of Tour Report*. Report to the Overseas Development Administration 15 December 1989. London.

Dupré, G., 1991, 'Les arbres, le fourré et le jardin: les plantes dans la société de Aribinda, Burkina Faso', in G. Dupré (ed.) *Savoirs paysans et développement*, 181–194. Paris: Karthala-ORSTOM.

Dupuis, J., 1824 (1966), *Journal of a Residence in Ashantee*, edited with notes and an introduction by W.E.F. Ward. London: Frank Cass.

Ebregt, A., 1995, *Report on Tropical Rainforest and Biodiversity Conservation in*

Ghana. The Netherlands: Dutch Government Ministry of Foreign Affairs Directorate General for International Cooperation.

Effah-Gyamfi, K., 1979, 'Dating potential of some plants in the West African savanna', post-graduate seminar papers, Department of History, Ahmadu Bello University, Zaria, Nigeria.

Ehui, S.K. and T.W. Hertel, 1989, 'Deforestation and agricultural productivity in the Côte d'Ivoire', *American Journal of Agricultural Economics* August: 703–711.

Ekanza, S.-P., 1981, 'Le Moronou à l'époque de l'administrateur Marchand: aspects physiques et économiques', *Annales d'Université d'Abidjan*, series I, History, 9: 55–70.

Ellis, J.E. and D.M. Swift, 1988, 'Stability of African pastoral ecosystems: alternate paradigms and implications for development', *Journal of Range Management* 41: 450–459.

Eltis, D., 1987, *Economic Growth and the Ending of the Transatlantic Slave Trade*. New York: Oxford University Press.

Elton, C., 1930, *Animal Ecology and Evolution*. Oxford: Clarendon Press.

EPC (Environment Protection Council), 1991, *Environment and Development in Ghana*. Report to the United Nations Conference on Environment and Development (UNCED). Accra: Environmental Protection Council.

Escobar, A., 1995, *Encountering Development: The Making and Unmaking of the Third World*. Princeton, New Jersey: Princeton University Press.

Escobar, A., 1996, 'Constructing nature: elements for a poststructural political ecology', in R. Peet and M. Watts (eds) *Liberation Ecologies: Environment, Development, Social Movements*. London and New York: Routledge.

Estève, J., 1983, 'La destruction du couvert forestier consécutive à l'exploitation forestière de bois d'Oeuvre en forêt dense tropicale humide africaine ou américaine', *Bois et Forêts des Tropiques* 201: 77–84.

Faillat, J.-P., 1990, 'Origine des nitrates dans les nappes de fissures de la zone tropicale humide – exemple de la Côte d'Ivoire', *Journal of Hydrology* 113: 231–264.

Fair, D., 1992, 'Africa's rain forests – retreat and hold', *Africa Insight* 22(1): 23–28.

Fairhead, J., 1989, *Food Security in North and South Kivu, Zaire*. Report to Oxfam, Oxford.

—— 1992, 'Indigenous technical knowledge and natural resources management in Sub-Saharan Africa: a critical overview', paper prepared for Social Science Research Council project on African Agriculture, Dakar, January 1992.

Fairhead, J. and M. Leach, 1994a, 'Contested forests: modern conservation and historical land use in Guinea's Ziama reserve', *African Affairs* 93: 481–512.

—— 1994b, 'Termites, society and ecology: perspectives from Mande and Central West Atlantic regions', paper presented to the African Studies Association UK Biennial Conference, University of Lancaster, 5–7 September 1994.

—— 1995a, 'False forest history, complicit social analysis: rethinking some West African environmental narratives', *World Development* 23(6): 1023–1036.

—— 1995b, 'Reading forest history backwards: the interaction of policy and local land use in Guinea, 1893–1993', *Environment and History* 1(1), March.

—— 1996a, *Misreading the African Landscape: Society and Ecology in a Forest–Savanna Mosaic*. Cambridge and New York: Cambridge University Press.

—— 1996b, 'Reframing forest history: a radical reappraisal of the roles of people and climate in West African vegetation change', in T.S. Driver and G.P. Chapman (eds) *Timescales and Environmental Change*, 169–195. London: Routledge.

—— 1996c, 'Enriching the landscape: social history and the management of transition ecology in the forest–savanna mosaic (Republic of Guinea)', *Africa* 66(1): 14–36.

FAO, 1953, *World Forest Resources*. Results of inventory undertaken in 1953. Rome: FAO.

—— 1981, *Forest Resources of Tropical Africa, Part I & II (Country Briefs)*. Tropical Forest Resources Assessment Project. Rome: FAO.

—— 1988, *Côte d'Ivoire. Programme Sectoriel forestier. Rapport de préparation*. Rapport du Programme de Coopération FAO/Banque Mondiale. Centre d'Investissement. Rome: FAO.

—— 1993, 'Forest resources assessment 1990: tropical countries', *FAO Forestry Paper* 112.

FAO/UNEP/GRID, 1980, *Global Environment Monitoring System Pilot Project on Tropical Forest Cover Monitoring: Benin–Cameroon–Togo Project Implementation: Methodology, Results and Conclusions*. Rome: FAO.

FDA/IDA, 1985, *Forest resources mapping of Liberia*, Report UTF/LIR/008/LIR by E.T. Hammermaster, Monrovia.

Ferguson, J., 1990, *The Anti-Politics Machine: 'Development', Depoliticization and Bureaucratic State Power in Lesotho*. Cambridge: Cambridge University Press.

Field, M.J., 1943, 'The agricultural system of the Manya-Krobo.' *Africa* 14(2): 54–65.

—— 1948, *Akim Kotoku and Oman of the Gold Coast*. London: Crown Agents for the Colonies.

Foggie, A., 1953, 'Forestry problems in the closed forest zone of Ghana', *Journal of the West African Science Association* 3: 131–147.

—— 1959, 'Forêts et foresterie au Ghana', *Bois et Forêts des Tropiques*, 65: 3–36.

Fontheneau, J. de, 1554 (1904), 'La cosmographie', in G. Musset (ed.) *Recueil des voyages et documents pour servir à l'histoire de la géographie*, vol. 20, Paris: Ernest Lerouse.

Ford, M., 1992, 'Kola production and settlement mobility among the Dan of Nimba, Liberia', *African Economic History* 20: 51–62.

Foresta, H. de, 1990, 'Origine et évolution des savanes intramayombiennes (R.P. du Congo). II. Apports de la botanique forestière', in R. Lanfranchi and D. Schwartz (eds) *Paysages quaternaires de l'Afrique Centrale atlantique*, 336–52. Paris: Didactiques, ORSTOM.

Foucault, M., 1980, *Power/Knowledge: Selected Interviews and Other Writings 1972–1977*. Edited by C. Gordon. Brighton: Harvester Press.

Fox, J.E.D., 1968, 'Exploitation of the Gola forest', *Journal of the West African Science Association* 13: 185–210.

Frank, W. and A. Gorgla, 1975, *Forest Inventory Report*. Bureau of Forestry, Ministry of Agriculture, Monrovia.

Freeman, R.A., 1898, *Travels and Life in Ashanti and Jaman*. London: Archibald Constable.

Freeman, T.B., 1844 (1968), *Journal of Various Visits to the Kingdoms of Ashanti, Aku and Dahomi in West Africa*. London: Cass (3rd edn of 1844 pub.).

Frimpong-Mensah, K., 1989, 'Requirement of the timber industry', in *Ghana Forest Inventory Proceedings*. Accra: Ghana Forest Department/ODA.

Fuller, F., 1921 (1967), *A Vanished Dynasty: Ashanti*. With a new introduction by W.E.F. Ward. 2nd edn (1st edn 1921). London: Cass.

Funtowicz, S.O. and Ravetz, J.R., 1992, 'Three types of risk assessment and the emergence of post-normal science', in S. Krimsky and D. Golding (eds) *Social Theories of Risk*, 251–273. Westport: Praeger.

Furley, P., J. Proctor and J. Ratter (eds), 1992, *Nature and Dynamics of Forest–Savanna Boundaries*. London: Chapman and Hall.

Fyfe, C., 1962, *A History of Sierra Leone*. Oxford: Oxford University Press.

Fyle, C.M., 1988a, 'North-east Sierra Leone in the nineteenth and twentieth centuries: reconstruction and population distribution in a devastated area', in C. Magbaily Fyle, *History and Socio-economic Development in Sierra Leone*, 33–60. Freetown: SLADEA.

—— 1988b, 'Population patterns, labour mobilization and agriculture in north-eastern Sierra Leone: a diachronic perspective', in C.M. Fyle (ed.) *History and Socio-economic Development in Sierra Leone: A Reader*, 197–212. Freetown: SLADEA.

Gaba, K.A., 1942, 'The history of Aneho ancient and modern'. Aneho, MS.

Gabel, C., R. Borden and S. White, 1974, 'Preliminary report on an archaeological survey of Liberia', *Liberian Studies Journal* 5: 87–105.

Garrett, G.H., 1892, 'Sierra Leone and the interior to the upper waters of the Niger', *Proceedings of the Royal Geographical Society* VII: 446.

Gayibor, N.L., 1986, 'Ecologie et histoire: les origines de la savane du Bénin', *Cahiers d'Etudes Africaines* 101–102, 24(1–2): 13–14.

—— 1988, 'Les origines de la savane du Bénin: une chasse gardée?', *Cahiers d'Etudes Africaines* 113, 29(1): 137–138.

Gent, J.R.P., 1925, 'Why protect our forests?' *Journal of the Gold Coast Agricultural and Commercial Society* 4(1): 46–51.

Gent, J.R.P. and H.W. Moor, 1927, 'Forestry conditions in Ho District, Togoland', *Empire Forestry Journal* 6(1): 238–251.

Geysbeek, T., 1994, 'A traditional history of the Konyan (15th–16th century): Vase Camara's epic of Musadu', *History in Africa* 21: 49–85.

Ghartey, K.K.F., 1989, 'Results of the inventory', in J.L.G. Wong (ed.) *Ghana Forest Inventory Project*. Seminar proceedings, 29–30 March 1989 (Accra): 32–46.

Gimpel, J., 1976, *The Medieval Machine*. New York.

Gold Coast Handbook, 1925, Accra: Government Press.

—— 1928, London: Crown Agents for the Government of the Gold Coast.

—— 1937, London: Crown Agents for the Government of the Gold Coast.

Gold Coast Survey Department, 1949, *Atlas of the Gold Coast*. Accra: Government Survey Department.

Gomez-Pompa, A. and A. Kaus, 1992, 'Taming the wilderness myth', *Bioscience* 24(4): 271–279.

Gordon, O.L.A., G. Kater and D.C. Schwarr, 1979, 'Vegetation and land use in Sierra Leone', UNDP/FAO Technical Report 2. AG: DP/SIL/73/002.

Gornitz, V. and NASA, 1985, 'A survey of anthropogenic vegetation changes in West Africa during the last century – climatic implications', *Climatic Change* 7: 285–325.

Goucher, C.L., 1981, 'Iron is iron 'til it is rust: trade and ecology in the decline of West African iron-smelting', *Journal of African History* 22: 179–189.

—— 1988, 'The impact of German colonial rule on the forests of Togo', in J. Richards and R.P. Tucker (eds) *World Deforestation in the Twentieth Century*. Durham and London: Duke Press Policy Studies.

Gourou, P., 1947, *Les pays tropicaux*. Paris.

Government of Sierra Leone, 1923, *Annual Report of the Lands and Forests Department 1923*. Freetown: Government Printers.

—— 1937, *Annual Report of the Lands and Forests Department 1937*. Freetown: Government Printers.

—— 1938, *Annual Report of the Lands and Forests Department 1938*. Freetown: Government Printers.

Grainger, A., 1993, *Controlling Tropical Deforestation*. London: Earthscan Publications.

—— 1996, 'An evaluation of FAO tropical forest resource assessment 1990', *The Geographical Journal* 162(1): 73–79.

Grégoire, J.M. and D. Gales, 1988, 'Comparaison de l'état du couvert végétal entre 1975 et 1985 sur le bassin du Niger en Guinée: exploitation de l'indice de végétation normalisé', *L'Agronomie Tropicale* 43(3): 177–184.

Grillo, R., 1997, 'Introduction', in R. Grillo and J. Stirratt (eds) *Discourses of Development: Anthropological Perspectives*. Oxford: Berg Press.

Gros, Cne., 1910, 'Mission forestière à la Côte d'Ivoire (1908–1909)', *Bulletin de la Société de Géographie Commerciale de Paris* 1910: 289–308.

Grove, R.H., 1994, 'Chiefs, boundaries and sacred groves: early nationalism and the defeat of colonial conservationism in Nigeria and the Gold Coast, 1870–1916', paper presented at the conference on 'Escaping Orthodoxy: environmental change assessments, local natural resource management and policy processes in Africa', Institute of Development Studies, Sussex, 13–14 September 1994.

—— 1995, *Green Imperialism: Colonial Expansion, Tropical Island Edens and the Origins of Environmentalism, 1600–1860*. Cambridge: Cambridge University Press.

Guelly, K.A., B. Roussel and M. Guyot, 1993, 'Initiation of forest succession in savanna fallows in SW Togo', *Bois et Forêts des Tropiques* 235: 37–48.

Guillaumet, J.L. and E. Adjanohoun, 1971, 'La végétation de la Côte d'Ivoire. Le milieu naturel de la Côte d'Ivoire', *Memoirs ORSTOM* 50: 156–263.

Guillot, B., 1980, 'La création et la destruction des bosquets Koukouya, symboles d'une civilisation et de son déclin', *Cahiers ORSTOM*, series Sciences Humaines, 17(3–4): 177–189.

Guyer, J. and P. Richards, 1996, 'The invention of biodiversity: social perspectives on the management of biological variety in Africa', *Africa* 66(1): 1–13.

Gyasi, E, G.T. Agyepong, E. Ardayfio-Schandorf, L. Enu-Kwesi, J.S. Nabila and E. Owusu-Bennoah, 1995, 'Production pressure and environmental change in the forest–savanna zone of Southern Ghana', *Global Environmental Change* 5(4): 355–366.

Haden-Guest, S., J.K. Wright and E.M. Teclaff (eds), 1956, *A World Geography of Forest Resources*, New York: Ronald Press Company.

Hair, P.E.H., 1962, 'An account of the Liberian hinterland c. 1780', *Sierra Leone Studies* 16: 218–226.

—— 1974, 'Barbot, Dapper, Davity: a critique of sources on Sierra Leone and Cape Mount', *History in Africa*, 1: 25–54.

—— 1978, 'Sources on early Sierra Leone (13). Barreira's report of 1607–1608. The visit to Bena', *Africana Research Bulletin* 8(2–3): 64–108.

—— 1979, 'Sources on early Sierra Leone (14). English accounts of 1582', *Africana Research Bulletin* 9 (1–2): 67–99.

Hall, C.A.S., Hanquin Tian, Ye Qi, G. Pontius, J. Cornell and J. Uhlig, 1995, 'Spatially explicit models of land use change and their application to the tropics', CDIAC (Oak Ridge National Laboratory) DOE Research Summary.

Hall, J., 1836, 'Tour of Dr. Hall up the Cavally river', *Missionary Herald* (USA) August: 312–314.

Hall, J.B., 1987, 'Conservation of forest in Ghana', *Universitas* 8: 33–42. University of Ghana at Legon.

Hall, J.B. and M.D. Swaine, 1976, 'Classification and ecology of closed canopy forest in Ghana', *Journal of Ecology* 64(3): 913–951.

—— 1981, 'Distribution and ecology of vascular plants in a tropical rain forest: forest vegetation in Ghana', *Geobotany* 1. The Hague: Dr W. Junk Publishers.

Hamilton, A.C., 1992, 'History of forests and climate', in J. Sayer, C.S. Harcourt and N.M. Collins (eds) *Conservation Atlas of Tropical Forests: Africa*, 17–25. London: Macmillan.

Harris, J.M., 1866, 'Some remarks on the origin, manners, customs and superstitions of the Gallinas people of Sierra Leone', *Memoirs read before the Anthropological Society of London* 1865–6: 25–36.

Hart, T.B., 1990, 'Monospecific dominance in tropical rainforests', *Tree* 5(1): 6–11.

Hasselmann, K.H., 1986, 'Liberian forests, geoecological ponderabilities', *Liberia Forum* 2/3: 26–60.

—— 1991, 'Problèmes économiques et écologiques des forêts denses libériennes', *Cahiers d'Outre-Mer* 44(1): 49–60.

Hawthorne, W.D., 1996, 'Holes and the sums of parts in Ghanaian forest: regeneration, scale and sustainable use', *Proceedings of the Royal Society of Edinburgh* 104B: 75–176.

Hawthorne, W. and A.J. Musah, 1995, *Forest Protection in Ghana with Particular Reference to Vegetation and Plant Species*. Gland, Switzerland and Cambridge, UK: IUCN in collaboration with ODA and the Forest Department, Republic of Ghana.

Hecht, S.B. and D.A. Posey, 1989, 'Preliminary results on soil management techniques of the Kayapo Indians', in D.A. Posey and W. Balée (eds) 'Resource management in Amazonia: indigenous and folk strategies', *Advances in Economic Botany* 7: 174–188.

Hellden, U., 1991, 'Desertification: time for an assessment?', *Ambio* 20(8): 372–383.

Henderson-Sellers, A. and V. Gornitz, 1984, 'Possible climatic impacts of land cover transformations, with particular emphasis on tropical deforestation', *Climatic Change* 6: 231–257.

Henige, D., 1986, 'Measuring the immeasurable: the Atlantic slave trade, West

African population and the Pyrrhonian critic', *Journal of African History* 27(2): 295–313.

Hervouet, J.-P. and C. Laveissière, 1987, 'Cash crop development and sleeping sickness in the forest belt of West Africa', in A. Akhtar (ed.) *Health and Disease in Tropical Africa: Geographical and Medical Viewpoints*, 373–381. London: Harwood.

Hill, M.H., 1972, 'Speculations on linguistic and cultural history in Sierra Leone', paper presented at the conference on Manding Studies, SOAS, London, 1972.

Hill, P., 1956, *The Gold Coast Cocoa Farmer*. Oxford: Oxford University Press.

—— 1963, *Migrant Cocoa Farmers of Southern Ghana*. Cambridge: Cambridge University Press.

Hills, T.L and R.E. Randall (eds), 1968, 'The ecology of the forest/savanna boundary', proceedings of the IGU Humid Tropics Commission Symposium, Venezuela, 1964. McGill University Savanna Research Project, *Savanna Research Series* 13.

Hirschmann, A.O., 1968, *Development Projects Observed*. Washington, DC: Brookings Institution.

Hoben, A., 1995, 'Paradigms and politics: the cultural construction of environmental policy in Ethiopia', *World Development* 23(6): 1007–1022.

Hoffman, C.C., 1862, 'A missionary journey up the Cavalha River, and the report of a large river flowing near the source of the former', *Proceedings of the Royal Geographical Society* 6, 1861–2, I–V: 66–67.

Holas, B. (1952) 'Mission dans l'est libérien: résultats démographiques, ethnologiques et anthropométriques', *Mémoire de l'IFAN* 14, Dakar.

Holling, C.S., 1973, 'Resilience and stability of ecological systems', *Annual Review of Ecology and Systematics* 4: 1–23.

Hollins, N.C., 1928, 'Mende law', *Sierra Leone Studies* 12: 25–29.

Holmberg, J. *et al.*, 1993, *Facing the Future: Beyond the Earth Summit*. London: IIED.

Holsoe, T., 1961, *Third Report on Forestry Progress in Liberia 1951–1959*. Washington, DC: International Cooperation Administration.

Holsoe, S.E. and J. Lauer, 1976, 'Who are the Kran/Guere? Ethnic identifications along the Liberia–Ivory Coast border', *African Studies Review* 19(1): 139–149.

Hopkins, B., 1962, 'Vegetation of the Olokomeji forest reserve, Nigeria I: general features and the research sites', *Journal of Ecology* 34: 20–87.

—— 1965 (1974), *Forest and Savanna*. London: Heinemann.

—— 1992, 'Ecological processes at the forest–savanna boundary', in P.A. Furley *et al.* (eds) *Nature and Dynamics of Forest–Savanna Boundaries*, 21–34. London: Chapman and Hall.

Houdaille, Cpt., 1900, 'Etude sur les propriétés et l'exportation des bois de la Côte d'Ivoire', *Revue des Cultures Coloniales* VI: 136.

Houghton, R.A., 1991, 'Tropical deforestation and atmospheric carbon dioxide', *Climatic Change* 19: 99–118.

Houghton, R.A., J.E. Hobbie, J.M. Melillo, B. Moore, B.J. Peterson, G.R. Shaver and G.M. Woodwell, 1983, 'Changes in the carbon content of terrestrial biota and soils between 1860 and 1980: a net release of CO_2 to the atmosphere', *Ecological Monographs* 53(3): 235–262.

Houghton, R.A., J.D. Unruh and P.A. Lefebvre, 1993, 'Current land cover in the tropics and its potential for sequestering carbon', *Global Biogeographical Cycles* 7(2): 305–320.

Houghton, R.A., J.L. Hackler and R.C. Daniels (eds), 1995, *Continental Scale Estimates of the Biotic Carbon Flux from Land Cover Change: 1850–1980*. Carbon Dioxide Information Analysis Center, Oak Ridge National Laboratory. Environmental Sciences Division Publication 4379.

Huguet, L., 1982, 'Que penser de la "disparition" des forêts tropicales?', *Bois et Forêts des Tropiques* 195: 7–22.

Hunter, J.M., 1972, 'Cocoa migration and patterns of land ownership in the Densu Valley, near Suhum, Ghana', in R.M. Prothero (ed.) *People and Land in Africa South of the Sahara*, 86. New York: Oxford University Press.

Hyde, W.F., G.S. Amacher and W. Magrath, 1996, 'Deforestation and forest land use: theory, evidence and policy implications', *The World Bank Research Observer* 11(2): 223–248.

Iroko, A.F., 1982, 'Le rôle des termitières dans l'histoire des peuples de la République Populaire du Bénin des origines à nos jours', *Bulletin de l'IFAN* 44, B1–2: 50–75.

Isert, P.E., 1788 (1989), *Voyages en Guinée et dans les Iles Caraibes en Amérique*. Translated into French, with introduction, notes and appendices by N. Gayibor. Paris.

—— 1788 (1992), *Letters on West Africa and the Slave Trade. Paul Erdmann Isert's Journey to Guinea and the Caribbean Islands in Columbia 1788*. Translated from the German and edited by S.A. Winsnes. British Academy: Oxford University Press.

IUCN, 1994, *Guidelines for Protected Area Management Categories*. IUCN Commission on National Parks and Protected Areas. Gland, Switzerland: IUCN.

IUCN/WWF/UNEP, 1991, *Caring for the Earth: A Strategy for Sustainable Living*. Gland, Switzerland: IUCN, WWF and UNEP.

Jaeger, P., M. Lamotte and R. Roy, 1966, 'Les richesses floristiques et faunistiques des monts Loma (Sierra Léone): urgence de leur protection intégrale', *Bulletin IFAN* 28: 1149–1190.

Jeffreys, M.W., 1950, 'Feux de brousse', *Bulletin IFAN* 3: 682–710.

Jepma, C.J., 1995, *Tropical Deforestation: A Socio-economic Approach*. London: Earthscan Publications.

Johnson, H.J., 1906, *Liberia*. London: Hutchinson.

Johnson, M., 1964, 'Migrant's progress', Part 1, *Bulletin of the Ghana Geographical Association* 9(2): 1–27.

—— 1965, 'Migrant's progress', Part 2, *Bulletin of the Ghana Geographical Association* 10(2): 13–40.

—— 1978, 'The population of Asante, 1817–1921: a reconsideration', *Asantesem* 8: 22–28.

—— 1981, 'Elephants for want of towns', in C. Fyfe and D. McMaster (eds) *African Historical Demography*. Volume II. Proceedings of a Seminar at the Centre of African Studies, Edinburgh, April 1981: 315–330.

Jones, A., 1981, 'Who were the Vai?', *Journal of African History* 22: 159–178.

—— 1983a, 'The Kquoja kingdom', *Paideuma* 29: 23–43.

—— 1983b, *From Slaves to Palm Kernels: A History of the Galinhas Country (West Africa) 1730–1890*. Wiesbaden: Franz Steiner Verlag.

Jones, A. and M. Johnson, 1981, 'Slaves from the windward coast', *Journal of African History* 22: 159–178.

Jones, E.W., 1956, 'The plateau forest of the Okomu forest reserve', *Journal of Ecology* 53: 54.

Jones, E.W., 1963, 'The forest outliers in the guinea zone of northern Nigeria', *Journal of Ecology* 51: 415–434.

Kahn, J. and J. McDonald, 1994, 'International debt and deforestation', in K. Brown and D.W. Pearce (eds) *The Causes of Tropical Deforestation: The Economic and Statistical Analysis of Factors Giving Rise to the Loss of Tropical Forests*. London: UCL Press.

Kalms, J.M., 1977, 'Studies of cultivation techniques at Bouaké, Ivory Coast', in D.J. Greenland and R. Lal (eds) *Soil Conservation and Management in the Humid Tropics*, 195–200. New York: John Wiley.

Kandeh, B. and P. Richards, 1996, 'Rural people as conservationists: querying neo-malthusian assumptions about biodiversity in Sierra Leone', *Africa* 66(1): 90–103.

Karl, E., 1974, *Traditions orales au Dahomey–Bénin.* Centre Regional de Documentation pour Tradition Orale, Niamey, Niger.

Keay, R.W.J., 1947, 'Notes on the vegetation of old Oyo reserve', *Farm and Forest*, January–June: 36–47.

—— 1959a, *Vegetation Map of Africa South of the Tropic of Cancer*. London: Oxford University Press.

—— 1959b, 'Derived savanna – derived from what?', *Bulletin IFAN*, series A2: 427–438.

—— 1973, 'Change in African vegetation', *African Research and Documentation* 1: 5–10.

Kershaw, A.P., 1992, 'The development of rainforest–savanna boundaries in tropical Australia', in P. Furley, J. Proctor and J. Ratter (eds) *Nature and Dynamics of Forest–Savanna Boundaries*, 255–272. London: Chapman and Hall.

Kiyaga-Mulindwa, D., 1982, 'Social and demographic changes in the Birim Valley, southern Ghana', *Journal of African History* 23: 63–82.

Klein, A. N., 1994, 'Slavery and Akan origins', *Ethnohistory* 41(4): 627–656.

—— 1996, 'Toward a new understanding of Akan origins', *Africa* 66(2): 248–273.

Klos-Kantowicz, E., 1973, 'The forest–savanna boundary against a background of differentiation of temperature and rainfall in Equatorial and West Africa', *Africana Bulletin (Warsaw)*: 143–159.

Kouassi, P.K., 1987, 'Etude comparative de la macrofaune endogée d'écosystèmes guinéens naturels et transformés de Côte d'Ivoire', thèse de 3e cycle, Université Nationale de Côte d'Ivoire.

Kuevi, D., 1975, 'Le travail et le commerce du fer au Togo avant l'arrivée des Européens', *Etudes Togolaises* NS 11 and 12: 24–43.

Kwamena Poh, M.A., 1973, *Government and Politics in the Akuapem State 1730–1850*. London: Longman.

Labat, J.B., 1730, *Voyage du chevalier des Marchais en Guinée, isles voisines et à Cayenne, fait en 1725, 1726 et 1727*. Paris: Saugrain.

Laing, A.G., 1825, *Travels in the Timannee, Kooranko, and Soolima Countries in Western Africa*. London: John Murray.

Lal, R., 1976, 'Soil erosion on alfisols in Western Nigeria II: effects of mulch rates', *Geoderma* 16: 377–387.

—— 1987, *Tropical Ecology and Physical Edaphology*. Chichester: John Wiley.

Lamb, A. F., 1942, 'The Kurmis of northern Nigeria', *Farm and Forest* 3: 187–192.

Lane-Poole, C.E., 1911, *Report on the Forests of Sierra Leone*. Freetown.

de Lange, S.M.H. and J.H. Neuteboom, 1992, *Report on the Identification Mission of Conservation of Tropical Rainforest in Ghana*. The Netherlands: Wageningen.

Lanly, J.P., 1969, 'La regression de la forêt dense in Côte d'Ivoire' *BFT* 127: 45–59.

—— 1982, 'Tropical forest resources', *FAO Forestry Paper* 30. Rome: FAO.

Latham, M. and M. Dugerdi, 1970, 'Contribution à l'étude de l'influence du sol sur la végétation au contact forêt–savane dans l'ouest et le centre de la Côte d'Ivoire', *Adansonia* 10: 553–576.

Laws, R.M., 1970, 'Elephants as agents of habitat and landscape change in East Africa', *Oikos* 21: 1–15.

Leach, G. and R. Mearns, 1988, *Beyond the Woodfuel Crisis: People, Land and Trees in Africa*. London: Earthscan Publications.

Leach, M., 1994, *Rainforest Relations: Gender and Resource Use Among the Mende of Gola, Sierra Leone*. International African Library. Edinburgh: Edinburgh University Press and Washington, DC: Smithsonian Institution.

Leach, M. and J. Fairhead, 1994a, 'The forest islands of Kissidougou: social dynamics of environmental change in West Africa's forest–savanna mosaic', report to ESCOR of the Overseas Development Administration, July 1994.

—— 1994b, 'Natural resources management: the reproduction of environmental misinformation in Guinea's forest–savanna transition zone', *IDS Bulletin* 25(2): 81–87.

—— 1995, 'Ruined settlements and new gardens: gender and soil ripening among Kuranko farmers in the forest–savanna transition zone', *IDS Bulletin* 26(1): 24–32.

Leach, M. and R. Mearns (eds), 1996, *The Lie of the Land: Challenging Received Wisdom on the African Environment*. Oxford: James Currey Publishers and New York: Heinemann.

Leach, M., R. Mearns and I. Scoones, 1996, 'Environmental entitlements: a framework for understanding the institutional dynamics of environmental change', IDS Discussion Paper 359. Brighton: Institute of Development Studies.

Ledant, J.-P., 1984–5, 'La réduction de biomasse végétale en Afrique de l'Ouest: aperçu général', *Annales de Gembloux* 90: 195–216; 1985, 91: 111–123.

Lena, P., 1979, *Transformation de l'espace rural dans le front pionnier du sud-ouest ivoirien*. Abidjan: ORSTOM.

Lézine, A.-M., 1989, 'Late Quaternary vegetation and climate of the Sahel', *Quaternary Research* 32: 317–334.

Lézine, A.-M. and J. Casanova, 1989, 'Pollen and hydrological evidence for the interpretation of past climates in tropical West Africa during the Holocene', *Quaternary Science Reviews* 8: 45–55.

Little, K., 1967, *The Mende of Sierra Leone: A West African People in Transition*. London: Routledge and Kegan Paul.

Lovejoy, P., 1982, 'The volume of the Atlantic slave trade: a synthesis', *Journal of African History* 23: 473–502.

—— 1983, *Transformations in Slavery: A History of Slavery in Africa*. Cambridge: Cambridge University Press.

Lovejoy, P., 1989, 'The impact of the Atlantic slave trade on Africa: a review of the literature', *Journal of African History* 30: 365–394.

Macaire, lt., 1900, ' La richesse forestière de la Côte d'Ivoire', *Revue cult col.* 4, January: 33–42.

MacArthur, R. and E. Wilson, 1967, *The Theory of Island Biogeography*. Princeton: Princeton University Press.

McCann, J., 1995, *People of the Plough*. New York: Heinemann.

McCleod, N.C., 1920, Statement prepared for the British Empire Forestry Conference, 1920.

McIntosh, S.K. and R.J. McIntosh, 1992, 'Cities without citadels: understanding urban origins along the middle Niger', in T. Shaw, P. Sinclair, B. Andah and A. Okpoko (eds) *The Archaeology of Africa: Food, Metals and Towns*, 622–641. London: Routledge.

McNeely, J.A., 1994, 'Lessons from the past: forests and biodiversity', *Biodiversity and Conservation* 3: 3–20.

McNeely, J.A. and D. Pitt (eds), 1985, *Culture and Conservation: The Human Dimension in Environmental Planning*. London and Sydney: Croom Helm.

Maley, J., 1987, 'Fragmentation de la forêt dense humide africaine et extension des biotypes montagnards au Quaternaire récent: nouvelles données polliniques et chronologiques. Implications paléoclimatiques et biogéographiques', *Palaeoecology of Africa* 18: 307–334.

—— 1996, 'The African rain forest: principal patterns of vegetation from Upper Cretaceous to Quaternary', *Proceedings of the Royal Society of Edinburgh* 104B: 31–73.

Mangenot, G., 1955, 'Etude sur les forêts des plaines et plateaux de la Côte d'Ivoire', *Etudes Eburnéennes* 4: 5–61.

Mangin, M., 1924, 'Une mission forestière en Afrique-Occidentale française'. *La Géographie* 42: 449–628.

Manning, P., 1992, 'The slave trade: the formal demography of a global system', in J.E. Inikori and S.L. Engerman (eds) *The Atlantic Slave Trade: Effects on Economies, Societies and Peoples in Africa, the Americas and Europe*, 117–141. Durham and London: Duke University Press.

Markham, R.H. and A.J. Babbedge, 1979, 'Soil and vegetation catenas on the forest–savanna boundary in Ghana', *Biotropica* 11(3): 224–234.

Marsh, G.P. (D. Cowenthal, ed.), 1864 (1965), *Man and Nature*. Cambridge, MA: Belknap Press.

Martin, C., 1991, *The Rainforests of West Africa: Ecology – Threats, Conservation*. Basel, Boston and Berlin: Birkhauser Verlag.

Massing, A., 1980, *The Economic Anthropology of the Kru (West Africa)*. Wiesbaden: Franz Steiner Verlag. Studien zur Kulturkunde 55.

Mather, A.S., 1990, *Global Forest Resources*. London: Belhaven Press.

Matthews, E. 1983, 'Global vegetation and land use. New high resolution data bases for climate studies', *J. Clim. Appl. Met.* 33(3): 474–487.

Mauny, R., 1956, 'Esmeraldo de Situ Orbis: côte occidentale d'Afrique du sud marocain au Gabon par Duarte Pacheco Pereira (vers 1506–1508)', Centro de Estudos da Guiné Portuguesa No. 19. Bissau.

Mayer, K., 1951, 'Forest resources of Liberia', *Agricultural Information Bulletin*, Forest Service USDA no. 67. Washington, DC.

Meave, J. and M. Kellman, 1994, 'Maintenance of rain forest diversity in riparian forests of tropical savannas: implications for species conservation during Pleistocene drought', *Journal of Biogeography* 21: 121–135.

Menaut, J.C. and J. Cesar, 1979, 'Structure and primary productivity of Lamto Savannas, Ivory Coast', *Ecology* 60(6): 1197–1210.

Menaut, J.C., J. Gignoux, C. Prado and J. Clobert, 1991, 'Tree community dynamics in a humid savanna of the Côte d'Ivoire: modelling the effects of fire and competition with grass and neighbours', in P. Werner (ed.) *Savanna Ecology and Management: Australasian Perspectives and International Comparisons*, 127–137. Oxford: Blackwell Scientific Publications.

Meniaud, J., 1922, *La forêt de la Côte d'Ivoire et son exploitation*. Pref. de M. le Gouverneur Antonetti. Introduction et considérations générales sur le pays et les habitants, de M. M. Larre. Les publications africaines. Paris: Larosse.

—— 1933, 'L'arbre et le forêt en Afrique noire', *Académie des Sciences Coloniales: Comptes Rendus Mensuels des Séances de l'Académie des Séances Coloniales: Communications*, vol. 14, 1929–1930.

Metzger, O.F., 1911, *Unsere alte Kolonie Togo* (Neudamm).

Meyer, K.M., 1951, 'Forest resources of Liberia, U.S. Dept. of Agriculture', *Agricultural Information Bulletin* 67.

Miege, J., 1955, 'Les savanes et forêts claires de Côte d'Ivoire', *Etudes Eburnéennes* 4: 62–81.

—— 1966, 'Observations sur les fluctuations des limites savanes–forêts en basse Côte d'Ivoire', *Annales de la Faculté des Sciences* 19: 149–166, Dakar.

Migeod, F.W.H., 1926, A *View of Sierra Leone*. London: Kegan Paul, Trench and Trubner.

Millington, A.C., 1985, 'Soil erosion and agricultural land use in Sierra Leone', PhD thesis, University of Sussex.

Millington, A., 1987, 'Environmental degradation, soil conservation and agricultural policies in Sierra Leone, 1895–1984', in D. Anderson and R. Grove (eds) *Conservation in Africa: People, Policies and Practice*, 29–248. Cambridge: Cambridge University Press.

Mitja, D., 1990, 'Influence de la culture itinérante sur la végétation d'une savane humide de Côte d'Ivoire', unpublished PhD thesis, University of Pierre et Marie Curie, Paris.

Mitja, D. and H. Puig, 1991, 'Essartage, culture itinérante et reconstitution de la végétation dans les jachères en savane humide de Côte d'Ivoire (Booro-Borotou, Touba)', in C. Floret and G. Serpantié (eds) *La Jachère en Afrique de l'Ouest*. Report of International Workshop, Montpellier, 2–5 December 1991. Paris: ORSTOM Editions.

Moloney, 1887, *Sketch of the Forestry of West Africa, with Particular Reference to its Present Principal Commercial Products*. London: Sampson Low.

Mondjannagni, A., 1969, Contribution à l'étude des paysages végétaux du Bas-Dahomey', *Annales de l'Université d'Abidjan*, series G, Tome 1, fasc. 2.

Monnier, Y., 1968, 'Les effets des feux de brousse sur une savane préforestière de Côte d'Ivoire', *Etudes Eburnéennes* IX.
—— 1979, Contribution à l'étude des rapports entre l'homme et les formations tropicales. L'exemple de l'Ouest africain du golfe de Guinée au fleuve Niger', thèse d'Etat.
Monnier, Y., 1980, 'Meningite cerebro-spinale, harmattan et deforestation', *Cahiers d'Outre-Mer* 130 (April–June): 103–122.
—— 1981, *La poussière et la cendre: paysages, dynamique des formations végétales et stratégies des sociétés en Afrique de l'Ouest.* Paris: Agence de Coopération Culturelle et Technique.
Monnier, Y. and Coll, 1974, 'Découverte aérienne de la Côte d'Ivoire', MS, Université d'Abidjan.
Moor, H.W., 1924, 'Forestry and its application to the Gold Coast', *Journal of the Gold Coast Agricultural and Commercial Society* 3(2): 80–83.
—— 1936, 'Deforestation in the Bissa cocoa area, Gold Coast', *The Malayan Forester* July: 1–4.
—— 1937, 'Vegetation and climate', *Empire Forestry Journal* 16: 200–214.
—— 1939, 'The influence of vegetation on climate in West Africa with particular reference to the protective aspects of forestry in the Gold Coast', Institute Paper 17, Imperial Forestry Institute, University of Oxford.
Morgan, W.B. and R.P. Moss, 1965, 'Savanna and forest in Western Nigeria', *Africa* 35(3): 286–293.
Mortimore, M., 1989, *Adapting to Drought: Farmers, Famines and Desertification in West Africa.* Cambridge: Cambridge University Press.
Mortimore, M.J., E.U. Essiet and S.P. Patrick, 1990, *The Nature, Rate and Effective Limits of Intensification in the Smallholder Farming System of the Kano Close-Settled Zone.* Ibadan: Federal Agricultural Coordinating Unit.
Morton, J.K., 1968, 'Sierra Leone', in I. Hedberg and O. Hedberg (eds) Conservation of vegetation in Africa south of the Sahara. *Acta Phytogeographica Suec.* 54: 72–74.
Mundt, R.J., 1987, *Historical Dictionary of the Ivory Coast.* London: Scarecrow Press.
Murphy, W.P. and C.H. Bledsoe, 1987, 'Kinship and territory in the history of a Kpelle chiefdom (Liberia)', in I. Kopytoff (ed.) *The African Frontier: The Reproduction of Traditional African Societies*, 121–148. Bloomington and Indianapolis: Indiana University Press.
Myers, N., 1980, *Conversion of Tropical Moist Forests.* Washington, DC: National Academy of Sciences.
—— 1988, 'Threatened biotas: "hot spots" in tropical forests', *The Environmentalist* 8: 187–208.
—— 1994, 'Tropical deforestation: rates and patterns' in K. Brown and D.W. Pearce (eds) *The Causes of Tropical Deforestation: The Economic and Statistical Analysis of Factors Giving Rise to the Loss of Tropical Forests.* London: UCL Press.
Nash, T.A.M., 1948, *Tsetse Flies in British West Africa: Part III, Sierra Leone.* London: HMSO for the Colonial Office.
Nicholson, S.E., 1979, 'The methodology of historical climate reconstruction and its application to Africa', *Journal of African History* 20(1): 31–49.
—— 1980, 'Saharan climates in historic times', in M.A.J. Williams and H. Faure

(eds) *The Sahara and the Nile: Quaternary Environments and Prehistoric Occupation in Northern Africa*, 173–200. Rotterdam: A.A. Balkema.

Norris, R., 1789, *Memoirs of the Reign of Bossa Ahadie, King of Dahomey, an inland country of Guiney, to which are added the author's journey to Ahomey* [sic], the capital, etc. London: Cass.

Noy-Meir, I., 1982, 'Stability of plant–herbivore models and possible applications to savanna', in B. Huntley and B.H. Walker (eds) *Ecology of Tropical Savannas*, 591–609. Berlin: Springer Verlag.

Nyerges, A.E., 1987, 'The development potential of the Guinea savanna: social and ecological constraints in the West African "Middle Belt"', in P.D. Litle and M.M. Horowitz (eds) *Lands at Risk in the Third World: Local Level Perspectives*. Boulder and London: Westview.

—— 1988, 'Swidden agriculture and the savannization of forests in Sierra Leone', unpublished PhD thesis, University of Pennsylvania.

—— 1989, 'Coppice swidden fallows in tropical deciduous forest: biological, technological and socio-cultural determinants of the secondary succession', *Human Ecology* 17: 379–400.

—— 1996, 'Ethnography in the reconstruction of African land use histories: a Sierra Leone example', *Africa* 66(1): 122–143.

O'Beirne, B., 1821, 'Journal, January to April 1821', in B.L. Mouser (ed.) 1979, *Guinea Journals: Journeys into Guinea-Conakry during the Sierra Leone Phase, 1800–21*. Washington, DC: University Press of America.

Odum, E.P., 1952, 'Relationships between structure and function in the ecosystem', *Japanese Journal of Ecology* 12: 108–118.

Ogilby, J., 1670, *Africa, being an accurate description . . .* London.

Okali, D.U.U., 1992, 'Sustainable use of West African moist forest lands', *Biotropica* 24(2b): 335–344.

Owusu, E.S.K., 1976, 'Dormaa oral traditions', MS, Institute of African Studies, University of Ghana at Legon.

Painter, T.M., 1991, 'Approaches to improving natural resource use for agriculture in Sahelian West Africa: a sociological analysis of the "Aménagement/Gestion des Terroirs Villageois" approach and its implications for non-government organizations', CARE Agriculture and Natural Resources Technical Report No. 3.

Paivinen, R.T.M. and R. Witt, 1988, 'Application of NOAA/AVHRR data for tropical forest cover mapping in Ghana', paper presented at IUFRO meeting, 29 August to 2 September 1988.

Palo, M., 1994, 'Population and deforestation', in K. Brown and D.W. Pearce (eds) *The Causes of Tropical Deforestation: The Economic and Statistical Analysis of Factors Giving Rise to the Loss of Tropical Forests*. London: UCL Press.

Parren, M.P.E. and N.R. de Graaf, 1995, *The Quest for Natural Forest Management in Ghana, Côte d'Ivoire and Liberia*. Wageningen: Wageningen Agricultural University.

Paulian, R., 1947, 'Observations écologiques en forêt de basse Côte d'Ivoire', Paris: Paul Lechevalier.

Pazzi, R., 1979, *Introduction à l'histoire de l'aire culturelle ajatado*. Université du Bénin, Institut National des Sciences de l'Education.

Pearce, D.W. and K. Brown, 1994, 'Saving the world's tropical forests', in K.

Brown and D.W. Pearce (eds) *The Causes of Tropical Deforestation: The Economic and Statistical Analysis of Factors Giving Rise to the Loss of Tropical Forests*. London: UCL Press.

Pearce, F., 1997, 'Lost forests leave West Africa dry', *New Scientist* 18 January: 15.

Peet, R. and M. Watts (eds), 1996, *Liberation Ecologies: Environment, Development, Social Movements*. London and New York: Routledge.

Pélissier, P. (ed.), 1980, 'L'arbre en Afrique tropicale: la fonction et le signe', *Cahiers ORSTOM*, series Sciences Humaines, 17(3–4).

Peltre, P., 'Le V Baoulé (Côte d'Ivoire centrale): heritage géomorphologique et paléoclimatique dans le trace du contact forêt–savane', *ORSTOM Travaux et documents* 80.

Pereira, Duarte Pacheco, 1506–1508 (1956), in R. Mauny (ed.) *Esmeraldo de Situ Orbis: côte occidentale d'Afrique du sud marocain au Gabon*. Bissau: Centro de Estudos da Guiné Portuguesa No. 19.

Perrot, C.H., 1974, 'Ano Asema: mythe et histoire', *Journal of African History* 15: 199–222.

—— 1976, 'De la richesse au pouvoir: les origines d'une chefferie du Ndenye (Côte d'Ivoire). Analyse critique de documents oraux', *Cahiers d'Etudes Africaines* XVI(61–62): 173–187.

—— 1982, *Les Anyi-Ndenye*. Abidjan and Paris: CEDA/Sorbonne:

Persson, R., 1974, 'World forest resources', *Research Notes* 17. Stockholm: Department of Forest Survey, Royal College of Forestry.

Peterken, G.F., 1996, *Natural Woodland: Ecology and Conservation in Northern Temperate Regions*. Cambridge: Cambridge University Press.

Petit-Maire, N. and J. Riser, 1987, 'Holocene palaeohydrography of the Niger', in J.A. Coetzee and E.M. Van Zinderen Bakker (eds) *Palaeoecology of Africa*, Vol. 18: 135–141. Rotterdam: Balkema.

Pfeiffer, V., 1988, *Agriculture au Sud-Bénin: passé et perspectives*. Paris: Editions L'Harmattan.

Phillips, J., 1974, 'Effects of fire in forest and savanna ecosystems of sub-Saharan Africa', in T.T. Koslowski and C.E. Ahlgren (eds) *Fire and Ecosystems*. New York: Academic Press.

Pierre, J., 1979, 'Contribution à la biogéographie de la région guinéenne. III, Le genre *Acraea fabricius* (Lépidoptère)' *C. R. Soc. Biogeogr.* 481: 73–79.

Pimbert, M. and J. Pretty, 1995, 'Parks, people and professionals: putting "participation" into protected area management', UNRISD Discussion Paper 57.

Plè, J., 1900, 'Délimitation avec le Togo', *La Géographie* 1900.

Pocknell, S. and D. Annalay, 1995, 'Report of the East Anglia expedition to the Loma mountains', unpublished report, University of East Anglia, Norwich.

Portères, R., 1950, 'Problème sur la végétation de la basse Côte d'Ivoire', *Bull. Soc. Bota. Franc.* 97: 153–156.

Posnansky, M., 1982, 'Archaeological and linguistic reconstruction in Ghana', in C. Ehret and M. Posnansky (eds) *The Archaeological and Linguistic Reconstruction of African History*, 256–289. Berkeley: University of California Press.

Prah, E.A., 1994, 'Sustainable management of the tropical high forest of Ghana', MS. London: Commonwealth Secretariat, September 1994.

Prevost, A.A., 1746, *Histoire générale des Voyages . . .* Paris, Didot.

Prinsley, R.T. (ed.), 1990, *Agroforestry for Sustainable Production: Economic Implications*. London: Commonwealth Science Council.

Pullan, R.A., 1974, 'Farmed parkland in West Africa', *Savanna* 3(2): 119–151.

Raison, J.P., 1988, 'Les "parcs" en Afrique: état des connaissances et perspectives de recherches', Document de Travail, Centre d'Etudes Africaines (EHESS), Paris.

Ramsay, J.M. and R. Rose-Innes, 1963, 'Some quantitative observations on the effects of fire on the guinea savanna vegetation of northern Ghana over a period of eleven years', *African Soils* 8: 41–86.

Rand, A.L., 1951, 'Birds from Liberia with a discussion for barriers between Upper and Lower Guinea subspecies', *Fieldiana Zool.* 32: 561–653.

Rattray, R.S., 1916, 'The iron workers of Akpafu', *Journal of the Royal Anthropological Institute* NS 19: 431–435.

Raven, P.H., 1987, *We're Killing our World: The Global Ecosystem in Crisis*. Chicago: The MacArthur Foundation.

Reade, W., 1870, 'Report on a journey to the upper waters of the Niger from Sierra Leone', *Proceedings of the Royal Geographical Society* 14(1–5): 185–188.

—— 1873, *The African Sketchbook*. London.

Repetto, R., 1988, *The Forest for the Trees? Government Policies and the Misuse of Forest Resources*. Washington, DC: World Resources Institute.

—— 1990, 'Deforestation in the tropics', *Scientific American* 262(4): 18–24.

Republic of Côte d'Ivoire, 1988, Annuaire retrospectif des statistiques agricoles et forestières 1900–1983. Direction de la Programmation de la Budgétisation et du Contrôle de Gestion, Ministère de l'Agriculture et des Eaux et Forêts.

Republic of Guinea, 1988, *Politique Forestière et Plan d'Action*. Plan d'Action Forestier Tropical 1988. Conakry.

Richard, J., 1972, *Le contact forêt–savane dans le centre-ouest ivoirien (Seguela-Vavoua): aspects et significations*. Abidjan: ORSTOM.

Richards, P., 1985, *Indigenous Agricultural Revolution: Ecology and Food Production in West Africa*. London: Hutchinson.

—— 1987, 'On the south side of the garden of Eden: creativity and innovation in Sub-Saharan Africa', MS, Dept of Anthropology, University College London.

—— 1990, 'Agriculture as a performance', in R. Chambers, A. Pacey and L.A. Thrupp (eds) *Farmer First: Farmer Innovation and Agricultural Research*. London: IT Publications.

—— 1993, 'Biodiversity and the dynamics of African anthropogenic landscapes: case studies from Upper Guinean forest formation', paper presented at the African Studies Association 36th Annual Meeting, Boston.

—— 1994, 'Answering Mr. Kaplan', address given at the 1994 African Studies Association (UK) biennial conference, September 1994.

—— 1996, *Fighting for the Rainforest: War, Youth, and Resources in Sierra Leone*. Oxford: James Currey.

Richards, P.W., 1952, *The Tropical Rain Forest*. Cambridge: Cambridge University Press.

—— 1973, 'The tropical rainforest', *Science* 1973: 58–67.

Robbins, C.B., 1978, 'The Dahomey Gap. A reevaluation of its significance as a faunal barrier to West African high forest mammals', *Bulletin of the Carnegie Museum of Natural History* 6: 168–174.

Robertson, G.A., 1819, *Notes on Africa, particularly those Parts which are situated between Cape Verd and the River Congo . . .* London: Sherwood, Neely and Jones.
Rodney, W., 1970, *A History of the Upper Guinea Coast 1545–1800*. New York: Monthly Review Press.
Roe, E., 1991, '"Development narratives" or making the best of blueprint development', *World Development* 19(4): 287–300.
Roe, E., 1995, 'Except Africa: postscript to a special section on development narratives', *World Development* 23(6): 1065–1070.
Rompaey, R.S.A.R. van, 1993, 'Forest gradients in West Africa: a spatial gradient analysis', doctoral thesis, Department of Forestry, Wageningen Agricultural University, The Netherlands.
Rossi, G., 1983, 'Evolution récente de l'environnement bio-climatique dans la région des plateaux'. Lomé: Mission française de coopération.
Sachtler, M., 1968, 'General report on national forest inventory in Liberia', Technical Report no. 1. German Forestry Mission to Liberia, Monrovia.
Sandford, S., 1995, 'Improving the efficiency of opportunism: new directions for pastoral development', in I. Scoones (ed.) *Living with Uncertainty*, 174–182. London: Intermediate Technology Publications.
Sargos, R., 1928, 'Rapport général sur les bois coloniaux', *Actes et Comptes-Rendus de l'Association Colonies-Sciences* 34: 91–99.
Sauer, C.O., 1925, 'The morphology of landscape', *University of California Publications in Geography* 2(2): 19–53.
Savage, Dr., 1839, 'Dr. Savage's journal', *The African Repository and Colonial Journal* 15(9): 155–160; 15(10): 166–171.
Savill, P.S. and J.E.D. Fox, 1967, *Trees of Sierra Leone*. Freetown: Government of Sierra Leone.
Sayer, J., C.S. Harcourt and N.M. Collins, 1992, *Conservation Atlas of Tropical Forests: Africa*. London: Macmillan.
Schantz, H.L. and C.F. Marbut, 1923, *The Vegetation and Soils of Africa*. New York.
Schiffer, Ct., 1910, 'Une mission industrielle et commerciale en Afrique-Occidentale française', *Bulletin de la Société de Géographie Commerciale de Paris* 1910: 770–827.
Schnell, R., 1949, 'Essai de synthèse biogéographique sur la région forestière d'Afrique occidentale', *Notes Africaines* October: 29–35.
—— 1950a, *La forêt dense. Introduction à l'étude botanique de la région forestière d'Afrique Occidentale*. Manuels Ouest-Africaines 1. Paris: P. Lechavalier.
—— 1950b, Remarques préliminaires sur les groupements végétaux de la forêt dense ouest-africaine. *Bulletin d'IFAN* 12: 297–314.
—— 1971, *Introduction à la Phytogéographie des Pays Tropicaux*. 2 vols. Paris: Gauthier-Villars.
Schwartz, A., 1971, *Tradition et changements dans la société Guéré.* Mémoire ORSTOM 52. Paris: ORSTOM.
Schwartz, D., 1992, 'Assèchement climatique vers 3000B.P. et expansion Bantu en Afrique-Centrale atlantique: quelques réflexions', *Bull. Soc. Géol. France* 163(3): 353–361.
Scoones, I. (ed.), 1995, *Living with Uncertainty: New Directions in Pastoral Development in Africa*. London: Intermediate Technology Publications.
Scoones, I. and Thompson, J. (eds), 1994, *Beyond Farmer First: Rural People's*

Knowledge, Agricultural Research, and Extension Practice. London: Intermediate Technology Publications.

Scott Elliot, G.F., 1893, 'Report on the district traversed by the Anglo-French boundary commission', Colonial Reports – Miscellaneous No. 3: Sierra Leone: 5–53. London: HMSO.

Seignobos, C., 1980, 'Des fortifications végétales dans la zone soudano-sahélienne (Tchad et Nord-Cameroun)', *Cahiers ORSTOM* series Sciences Humaines, XVII(3–4): 191–222.

—— 1983, 'Végétations anthropiques dans la zone soudano-sahélienne: la problématique des parcs', *Cameroon Geographical Review*, 1–23.

Seymour, G.L., 1856, 'Journal of a tour into the country interior of Grand Bassa', *African Repository* 32(6): 169–182.

—— 1858, 'Borwandow's town, Pessey', letter 1 April 1858. *African Repository* 45: 245–249.

—— 1860, 'The journal of the journey of George L. Seymour to the interior of Liberia: 1858', *New-York Colonization Journal* 105, 108, 109, 111, 112.

Sharma, N. and R. Roe, 1992, 'Managing the world's forests', *Finance and Development* 29(2): 31–33.

Sharpe, B., 1996, 'First the forest. . . . Settlement history, conservation and visions of the future in SW Cameroon', paper presented at the conference on 'Contested Terrain: West African Forestry Relations, Landscapes and Processes', University of Birmingham, April 1996.

Shaw, T., P. Sinclair, B. Andah and A. Okpoko (eds), 1992, *The Archaeology of Africa: Food, Metals and Towns*. London: Routledge.

Shepherd, G., 1990, *Communal Management of Forests in the Semi-arid and Sub-humid Zones of Africa.* Report prepared for FAO Forestry Department. London: Overseas Development Institute.

Siddle, D.J., 1968, 'War towns in Sierra Leone: a study in social change', *Africa* 38: 47–56.

—— 1969, 'The evolution of rural settlement forms in Sierra Leone circa 1400 to 1968', *Sierra Leone Geographical Journal* 13: 33–44.

—— 1970, 'Location theory and the subsistence economy: the spacing of rural settlements in Sierra Leone', *Journal of Tropical Geography* 21: 79–90.

Sims, J., 1860, 'Scenes in the interior of Liberia: being a tour through the countries of Dey, Goulah, Pessah, Barlain, Kpellay, Suloang and King Boatswain's Tribes, in 1958', *New York Colonization Journal* 105, 108, 109, 111, 112.

Small, Donn., 1953, 'Some ecological and vegetational studies in the Gola Forest Reserve, Sierra Leone', MSc dissertation, Queen's University, Belfast.

Smith, W., 1744, *A New Voyage to Guinea.* London.

Sobey, D.G., 1978, 'Anogeissus groves on abandoned village sites in the Mole National Park, Ghana', *Biotropica* 10(2): 87–99.

Sodefor, 1975, *La forêt dense humide en Côte d'Ivoire.* Sodefor Report, Abidjan.

Sournia, G., 1979, 'Une richesse naturelle en voie de disparition', *Cahiers d'Outre-Mer* 127, July–Sept.

Spears, J., 1986, 'Côte d'Ivoire: key forest policy issues for the coming decade in the rain forest zone', Forest Subsector Discussion Paper, May 1986. Washington, DC: World Bank.

Spichiger, R. and C. Blanc-Pamard, 1973, 'Recherches sur le contact forêt–savane en Côte d'Ivoire: étude du recru forestier sur des parcelles cultivées en lisière d'un ilôt forestier dans le sud du pays baoulé', *Candollea* 28: 21–37.

Spichiger, R. and V. Lassailly, 1981, 'Recherches sur le contact forêt–savane en Côte d'Ivoire: note sur l'évolution de la végétation dans la région de Béoumi (Côte d'Ivoire centrale) , *Candollea* 36: 145–153.

Sprugel, D.G., 1991, 'Disturbance, equilibrium, and environmental variability: what is "natural" vegetation in a changing environment?', *Biological Conservation* 58: 1–18.

Stahl, A.B., 1992, 'Intensification in the West African late stone age: a view from central Ghana', in T. Shaw, P. Sinclair, B. Andah and A. Okpoko (eds) *The Archaeology of Africa: Food, Metals and Towns*. London: Routledge.

Starr, F. (ed.), 1912, *Narrative of the Expedition Despatched to Musahdu by the Liberian Government under Benjamin K. Anderson Esq. in 1874*. Monrovia: College of West Africa Press.

Stibig, H.-J. and R. Baltaxe, 1993, 'Use of NOAA remote sensing data for assessment of the forest area of Liberia', RSC Series 66, Rome: FAO.

Stocking, M., 1996, 'Soil erosion: breaking new ground', in M. Leach and R. Mearns (eds) *The Lie of the Land: Challenging Received Wisdom on the African Environment*, 140–154. Oxford: James Currey.

Strong, R.P. (ed.), 1930, *Harvard African Expedition, The African Republic of Liberia and the Belgian Congo*. Cambridge, MA: Harvard University Press.

Surgey, A. de., 1988, *Le système religieux des Evhe*. Paris: L'Harmattan.

—— 1994, *Nature et fonction des fétiches en Afrique noire*. Paris: L'Harmattan.

Swaine, M.D. and J.B. Hall, 1988, 'The mosaic theory of forest regeneration and the determination of forest composition in Ghana', *Journal of Tropical Ecology* 4(3): 253–269.

Swaine, M.D., J.B. Hall and J.M. Lock, 1976, 'The forest–savanna boundary in west-central Ghana', *Ghana Journal of Science* 16(1): 35–52.

Swift, J., 1996, 'Desertification: narratives, winners and losers', in M. Leach and R. Mearns (eds) *The Lie of the Land: Challenging Received Wisdom on the African Environment*, 73–90. Oxford: James Currey.

Talbot, M.R., 1981, 'Holocene changes in tropical wind intensity and rainfall: evidence from southeast Ghana', *Quaternary Research* 16: 201–220.

Talbot, M.R. and G. Delibrias, 1977, 'Holocene variations in the level of Lake Bosumtwi, Ghana', *Nature* 268: 722–724.

Talbot, M.R., D.A. Livingstone, P.G. Palmer, J. Maley, J.M. Melack, G. Delibrias and S. Gulliksen, 1984, 'Preliminary results from sediment core from Lake Bosumtwi, Ghana', in J.A. Coetzee and E.M. Van Zinderen Bakker (eds) *Palaeoecology of Africa*, Vol. 16: 173–193. Rotterdam: Balkema.

Tanzidani, T.K.T., 1993, 'Les problèmes sociaux dans les réserves de faune du Togo', *Cahiers d'Outre-Mer* 46(181): 61–73.

Taylor, C.J., 1952, *The Vegetation Zones of the Gold Coast*. Accra: Government Printing Department.

—— 1960, *Synecology and Silviculture in Ghana*. London: Nelson.

Tereau, Lt. Col. and Dr Huttel, 1949, 'Monographie du Hollidge', *Etudes Dahoméennes* 11.

Thomann, G.. 1902–3. 'Mission de Sassandra à Séguéla', *Revue Coloniale* 1902/3: 621–653.

—— 1904, *A la Côte d'Ivoire. La Sassandra*. Imp. du Journal des Debats, 1904.

—— 1905, *Essai de manuel de la langue néouole, parlée dans la partie occidentale de la Côte d'Ivoire. Ouvrage accompagné d'un recul de contes et chansons en langue néoule, d'une étude sur les diverses tribus Beté-Bakoué, de vocabulaires comparatifs, d'une bibliographie et d'une carte*. Paris: Ernst Leroux.

Thomas, A.S., 1942, 'A note on the distribution of *Chlorophora excelsa* in Uganda', *Empire Forestry Journal* 21: 42–43.

Thomas, D.G., 1924, 'Report by the Forest Authority, Sierra Leone to the British Empire Forestry Conference', Sessional Paper 7 of 1924. Freetown: Government Printing Office and London: Crown Agents.

Thompson, H., 1910, *Gold Coast: Report on Forests*. Colonial Reports – Miscellaneous No. 66. London: HMSO.

—— 1911, 'The forests of southern Nigeria', *Journal of the African Society* 10(38): 120–145.

Thompson, M., 1993, 'Good science for public policy', *Journal of International Development* 5(6): 669–679.

Thulet, J.-Ch., 1981, 'La disparition de la forêt ivoirienne: pertes et profits pour une société', *L'Information Géographique* 45: 153–160.

Tiffen, M., Mortimore, M. and Gichuki, F., 1994, *More People, Less Erosion: Environmental Recovery in Kenya*. Chichester: John Wiley.

Trotter, J.K., 1897, 'An expedition to the source of the Niger', *The Geographical Journal* 10: 386–401.

Unwin, A.H., 1909, *Report on the Forests and Forestry Problems in Sierra Leone*. London: Waterlow.

—— 1920, *West African Forests and Forestry*. London: Fisher Unwin.

USAID, 1980, 'Phase I environmental profile of Liberia', US Agency for International Development, Office of Science and Technology.

Vansina, J., 1985, 'L'Homme, les forêts et le passé en Afrique', *Annales, Economie, Sociétés, Civilisations* 6: 1307–1334.

Viard, E.R., 1934a, 'Le Cercle de Guiglo (Côte d'Ivoire)', *La Géographie* 60: 100–116.

—— 1934b, 'Le Cercle de Guiglo (Côte d'Ivoire)', *La Géographie* 61: 232–269.

—— 1934c, *Les Guéré: peuple de la forêt*, Paris: Société d'éditions géographiques, maritimes et coloniales.

Vigne, C., 1937, Letter to the editor, *Empire Forestry Journal* 16: 93–94.

Villaut, N., 1669, *Relation des costes d'Afrique appellée Guinée*. Paris.

Visser, L., 1977, L'Igname, bonne à manger et bonne à penser: quelques aspects de l'agriculture Ahouah (Côte d'Ivoire)', *Cahiers d'Etudes Africaines* XVII(4): 535–544.

Voorhoeve, A.G., 1964, 'Some notes on the tropical rainforest of the Yoma-Gola National Forest near Bomi Hills, Liberia', *Commonwealth Forestry Review* 43(1): 17–24.

—— 1979, 'Liberian high forest trees. A systematic botanical study of the 75 most important or frequent high forest trees, with reference to numerous related species'. Wageningen: Centre for Agricultural Publishing. Report 652.

Vuattoux, R., 1970, 'Observations sur l'évolution des strates arborées et arbustives dans la savane de Lamto', *Annales de l'Université d'Abidjan*, series E, 3(1): 285–315.

Waldock, E.A., E.S. Capstick and A.J. Browning, 1951, 'Soil conservation and land use in Sierra Leone', Sessional paper 1 of 1951. Freetown: Government Printer.

Ward, W.E.F., 1958, *A History of Ghana*. London: Allen and Unwin.

Warnier, J.P. and I. Fowler, 1979, 'A nineteenth-century Ruhr in Central Africa', *Africa* 49(4): 329–350.

Warren, D.M., L.J. Slikkerveer and D. Brokensha (eds), 1995, *The Cultural Dimension of Development: Indigenous Knowledge Systems*. London: Intermediate Technology Publications.

Wartena, D., 1992, 'Comme le monde évolué: une histoire agronomique du plateau d'Aplahoue, Bénin, 1894–1986'. Mémoire de maîtrise en histoire rurale, Wageningen, Département d'Histoire Rurale.

—— 1994, 'Zunko, ordure ou engrais? La fumure des déchets ménagers dans six villages Adja et six villages Fon', in P. Ton and L. de Haan (eds) *A la recherche de l'agriculture durable au Bénin*, 75–87. Amsterdamse Sociaal-geografische Studies 49, Instituut voor Sociale Geografie, Universitat van Amsterdam.

Washington, R., 1994, 'The predictability of tropical African rainfall', paper presented at the African Studies Association UK biennial conference, Lancaster, 3–7 September 1994.

Watt, J., 1794, *Journal of James Watt: Expedition to Timbo capital of the Fula Empire in 1794*. B.L. Mouser (ed.) 1994. African Studies Program, University of Wisconsin-Madison.

Weiskel, T.C., 1980, *French Colonial Rule and the Baoulé Peoples, 1889–1911*. Oxford, Clarendon Press.

Wells, M., K. Brandon and L. Hannah, 1992, *People and Parks: Linking Protected Area Management with Local Communities*. Washington, DC: World Bank/WWF/ USAID.

Westoby, M., B. Walker and I. Noy-Meir, 1989, 'Opportunistic management for rangelands not at equilibrium', *Journal of Range Management* 42: 266–274.

White, F., 1983, *The Vegetation of Africa*. Paris: UNESCO.

Whitmore, T.C., 1990, *An Introduction to Tropical Rain Forests*. Oxford: Clarendon Press.

Wilkie, D.S., 1994, 'Remote sensing imagery for resource inventories in central Africa: the importance of detailed field data', *Environmental Economics* 1994: 379–403.

Wilks, A.A., 1978–9, 'Huppenbauer's account of Kumasi in 1881', *Asantesem* 8: 50–52; 9: 58–62; 10: 59–63.

Wilks, I., 1975, *Asante in the Nineteenth Century: The Structure and Evolution of a Political Order*. Cambridge: Cambridge University Press.

—— 1977, 'Land, labour, capital and the forest kingdom of Asante: a model of early change', in J. Friedman and M.J. Rowlands (eds) *The Evolution of Social Systems*, ed. London: Duckworth.

—— 1978, 'The population of Asante, 1817–1921: a rejoinder', *Asantesem* 8: 28–35.

—— (ed.), 1993, *Forests of Gold: Essays on the Akan and the Kingdom of Asante*. Athens, OH: Ohio University Press.

Willans, R.H.K., 1909, 'The Konnoh people', *Journal of the Africa Society* 8: 130–144, 288–295.

Wills, J.B. (ed.), 1962, *Agriculture and Land Use in Ghana*. London: Oxford University Press for Ghana Ministry of Food and Agriculture.

Wilson, E.O., 1989, 'Threats to biodiversity', *Scientific American* 261: 108–116.

Wilson, J.L., 1836, 'Extracts from the journal of Mr. Wilson', *Missionary Herald*, May: 193–197; June: 242–248; July: 387.

Wilson, L.E., 1991, *The Krobo People of Ghana to 1892: A Political and Social History*. Ohio University Center for International Studies, Monographs in International Studies, Africa Series no. 58.

Wilson, R.T., 1988, 'Vital statistics of the baobab (*Adansonia digitata*)', *African Journal of Ecology* 26: 197–206.

Winsers, S.A., 1992, 'P.E. Isert in German, French and English: a comparison of translations', *History in Africa* 19: 401–410.

Winterbottom, T.M., 1803 (1965), *An Account of the Native Africans in the Neighbourhood of Sierra Leone*. Reprint 1965. London: Frank Cass.

Wondji, C., 1963, 'La Côte d'Ivoire occidentale: période de pénétration pacifique (1890–1908)', *Revue Française d'Histoire d'Outre-Mer* 50: 346–381.

Wood, D., 1993, 'Forests to fields: restoring tropical lands to agriculture', *Land Use Policy* April: 91–107.

World Bank, 1988, Ghana Forest Resource Management Project. Working Papers 1–6. Unpublished.

—— 1991, *The Forest Sector: A World Bank Policy Paper*. Washington, DC: World Bank.

Worster, D., 1993, *The Wealth of Nature: Environmental History and the Ecological Imagination*. Oxford: Oxford University Press.

—— 1994, 'Nature and the disorder of history', *Environmental History Review* 18(2): 1–16.

Wynne, B., 1992, 'Uncertainty and environmental learning: reconceiving science and policy in the preventive paradigm', *Global Environmental Change* 2(2): 111–127.

Zheng, X. and E.A.B. Eltahir, 1997, 'The response to deforestation and desertification in a model of West African monsoons', *Geophysical Research Letters* 24(2): 155–158.

Zimmermann, J., 1867, 'Der Kroboneger', *Der evangelische Heidenboten* 36.

Zoller, H., 1885 (1989), *Le Togo et la Côte d'Esclaves* (trans. by K. Amegan and A. Ahadji). Paris/Lomé: ORSTOM-Karthala.

Zon, R. and W.N. Sparhawk, 1923, *Forest Resources of the World*. New York: McGraw-Hill.

Zweifel, J. and M. Moustier, 1879, *Voyage aux sources du Niger*. Marseille: Barlatier-Feissat.

INDEX